The Basics of Cellular

IEC
Chicago, Illinois

All opinions expressed in *The Basics of Cellular* are those of the authors and are not
binding on the International Engineering Consortium or Professional Education
International.

Library of Congress Cataloging-in-Publication Data

Chandler, Nick.
  The basics of cellular / Nick Chandler, Stephan S. Jones.
      p. cm.
  Includes bibliographical references.
  ISBN-13: 978-1-931695-49-7
  1. Mobile communication systems. 2. Wireless communication systems.
I. Jones, Stephan. II. Title.
  TK6570.M6C485 2006
  621.3845'6--dc22

                                                              2006008070

ISBN: 978-1-931695-49-7

International Engineering Consortium
300 West Adams Street, Suite 1210
Chicago, Illinois 60606-5114, USA
+1-312-559-4100 voice • +1-312-559-4111 fax
publications@iec.org • *www.iec.org*

About the International Engineering Consortium

The International Engineering Consortium (IEC) is a non-profit organization dedicated to catalyzing technology and business progress worldwide in a range of high technology industries and their university communities. Since 1944, the IEC has provided high-quality educational opportunities for industry professionals, academics, and students. In conjunction with industry-leading companies, the IEC has developed an extensive, free on-line educational program. The IEC conducts industry-university programs that have substantial impact on curricula. It also conducts research and develops publications, conferences, and technological exhibits that address major opportunities and challenges of the information age. More than 70 leading high-technology universities are IEC affiliates, and the IEC handles the affairs of the Electrical and Computer Engineering Department Heads Association and Eta Kappa Nu, the honor society for electrical and computer engineers. The IEC also manages the activities of the Enterprise Communications Consortium.

About the Authors

Nick Chandler is a Technical Support Engineer at ADTRAN, where he has supported an array of enterprise voice and data network products. Most recently, he has worked with the company's VoIP PBX product line, as well as LAN switches. His current areas of research include network security, wireless applications, and packet-voice media and control protocols. He holds a Bachelor of Arts degree in Telecommunications and a Master of Science degree in Information & Communication Sciences from Ball State University.

Stephan S. Jones, Ph.D., is a Professor at the Center for Information and Communication Sciences at Ball State University. Prior to coming to higher education, he was the owner and applications engineer for one of the country's largest interconnect companies providing telecommunication solutions to commercial clients. He has implemented local- and wide-area voice and data networks as well as designing distance-learning communities for higher education applications. He holds undergraduate degrees in Psychology and Engineering Technology. His Master's and Doctoral studies were done with cognate areas in Technology Education. His current research interests are in wireless application technologies, competitive broadband access technologies, secondary and tertiary market access to broadband services, and qualitative research applied to technical environments.

The IEC's University Program, which provides grants for full-time faculty members and their students to attend IEC Forums, is made possible through the generous contributions of its Corporate Members. For more information on Corporate Membership or the University Program, please call +1-312-559-4625 or send an e-mail to cmp@iec.org.

Based on knowledge gained at IEC Forums, professors create and update university courses and improve laboratories. Students directly benefit from these advances in university curricula. Since its inception in 1984, the University Program has enhanced the education of more than 500,000 students worldwide.

IEC Affiliated Universities

The University of Arizona
Arizona State University
Auburn University
University of California at Berkeley
University of California, Davis
University of California, Santa Barbara
Carnegie Mellon University
Case Western Reserve University
Clemson University
University of Colorado at Boulder
Columbia University
Cornell University
Drexel University
École Nationale Supérieure des Télécommunications de Bretagne
École Nationale Supérieure des Télécommunications de Paris
École Supérieure d'Électricité
University of Edinburgh
University of Florida
Georgia Institute of Technology

University of Glasgow
Howard University
Illinois Institute of Technology
University of Illinois at Chicago
University of Illinois at Urbana/Champaign
Imperial College of Science, Technology and Medicine
Institut National Polytechnique de Grenoble
Instituto Tecnológico y de Estudios Superiores de Monterrey
Iowa State University
KAIST
The University of Kansas
University of Kentucky
Lehigh University
University College London
Marquette University
University of Maryland at College Park
Massachusetts Institute of Technology
University of Massachusetts

McGill University
Michigan State University
The University of Michigan
University of Minnesota
Mississippi State University
The University of Mississippi
University of Missouri-Columbia
University of Missouri-Rolla
Technische Universität München
Universidad Nacional Autónoma de México
North Carolina State University at Raleigh
Northwestern University
University of Notre Dame
The Ohio State University
Oklahoma State University
The University of Oklahoma
Oregon State University
Université d'Ottawa
The Pennsylvania State University

University of Pennsylvania
University of Pittsburgh
Polytechnic University
Purdue University
The Queen's University of Belfast
Rensselaer Polytechnic Institute
University of Southampton
University of Southern California
Stanford University
Syracuse University
University of Tennessee, Knoxville
Texas A&M University
The University of Texas at Austin
University of Toronto
VA Polytechnic Institute and State University
University of Virginia
University of Washington
University of Wisconsin-Madison
Worcester Polytechnic Institute

Table of Contents

Chapter I: Introduction

With each passing year, communication systems become more prevalent in our world. There is no question that people are much more connected than ever before. Whether for business or personal use, immediate access to information and communication resources has become a part of day-to-day life for many people. Perhaps more than any other technology, mobile phones have become an integral part of this trend.

This book is designed to provide an overview of the essential elements of cellular networks, from mobile phones to antennas to protocols in use in these systems today (and everything in between). As the name of the book suggests, we will focus on the basics of these technologies—no math or physics degrees are required! Cellular networking topics are, admittedly, not always easy to understand, but the reader will gain insight into the theory and implementation of this increasingly popular means of communication. Because many of these issues also arise in other wireless systems, much of the material could also be applied in whole or in part to technologies such as wireless local area networks (WLANs).

Before we begin to explore mobile wireless systems in depth in *Chapter 2*, we will take a look at the history of wireless telecommunication systems as well as the regulatory factors surrounding such systems. We will also examine some of the basic principles that are essential to all wireless communication.

Historical Perspectives

The field of telecommunications began in the mid-1800s with the invention of the telegraph. Samuel F. B. Morse, who is most often remembered as the developer of Morse Code, invented the first electric telegraph in 1844. In May of that year, the first intercity telegraph was sent from the Capitol Building in Washington to Baltimore, Maryland. This first telegraph read, "What Hath God Wrought!"—in light of the advances that have taken place in the more than 150 years since, this phrase seems particularly fitting. Telegraphy became a popular means of communication soon thereafter and was, in fact, a key tool for military communications in the Civil War.

Thirty-two years after the first telegraph, another invention came along that was bound to make the telegraph obsolete. Although a number of inventors had been working independently to develop the technology to transmit speech electrically, Scottish-born Alexander Graham Bell succeeded in 1876. Initially, telephones were connected point-to-point; therefore, if a person needed to be able to call three people, he would need three different phones! Bell quickly saw the problems inherent in this system, and he implemented the first telephone exchange in 1878, which allowed calls to be switched on a per-call basis. By 1884 New York and Boston were connected via long-distance trunks. Bell stands out as one of the most important individuals in the history of communications. Not only did he invent the telephone and do much of the earliest development on it, but he also created the company that became AT&T. (In all fairness, however, the growth of the company was a result of the work of a number of people—in particular, Theodore Vail was essential to the company's early success.) For almost 100 years, AT&T was the telecommunications industry!

For more than twenty years after Bell's original telephone, telecommunications was dependent on physical connections. In the late 1890s, however, an Italian engineer, Guglielmo Marconi, produced a series of inventions (based on the work of physicists Joseph Henry and Michael Faraday) that proved wireless communication was practical. In 1899 Marconi successfully established wireless communication between France and England over the

English Channel. Somewhat more spectacularly, he produced the first trans-Atlantic wireless link in 1902 (the distance was more than 2000 miles). This early work set in motion the research and development that has led to the current wireless technologies that surround us today.

Mobile telephony was introduced to the world in St. Louis, Missouri, in 1946. The Mobile Telephone Service (MTS) of AT&T/Southwestern Bell allowed customers to use car phones for communication. Compared to newer systems, the MTS was very primitive. Very few users could access the network at once because a single antenna was used to transmit to all the mobile phones in the city. Conversely, multiple receivers were scattered throughout the city to receive signals from the mobile users and transmit back toward the public phone system. In addition only one party could talk at a time; mobile users had to hold a button when they wanted to talk and release the button to listen to the other party, much like walkie-talkies. The MTS did not provide a method for mobile users to directly dial phone numbers—instead, users had to page an operator to have calls connected. In 1948, The Richmond Radiotelephone Company, which was not associated with the Bell system, offered the first operator-free mobile service in Richmond, Indiana. For a number of years, mobile service could not be considered "cellular." As we will see in *Chapter 2*, cellular networks operate most efficiently when numerous antennas and radio transceivers are combined in a fashion that creates a series of cells. Radio frequencies in one cell may be reused in other cells without interference, thereby allowing a greater number of users to access the system. By the early 1980s, however, truly cellular networks were put into place, and in 1983 the first commercial cellular call was completed in Chicago. At the time, AT&T served Chicago, and in 1984 a competitor, Motorola/American Radio Telephone Service (ARTS), opened service up in the Washington, D.C., area.

Since the 1980s the popularity of cellular services has increased exponentially. Electronic components have improved and shrunk to the point that cellular phones can easily fit in even the smallest pants pockets. At the same time cellular providers now offer much more than just phone service and, as a result of advances in transmission technology, users can send pictures, videos, and instant messages to one another; furthermore, they can even access their e-mail or the World Wide Web from their cellular phones! Currently it appears that in the foreseeable future, cellular offerings will only be limited by the creativity of its innovators.

Regulatory Overview

The first major thrust of wireless regulation in America came in 1912, largely as a response to the sinking of the Titanic. Legislation from 1910 required that passenger ships above a certain size have radios onboard, which could be used for distress calls in the event of an emergency. As the Titanic began to sink in April 1912, its radio operator did, in fact, generate distress calls on the ship's radio system. Meanwhile, another ship, known as the California, floated approximately 20 miles away. As per the 1910 legislation, the California had a radio onboard. Unfortunately, the legislation did not require that ships leave their radios on at all times—as fate would have it, the California's radio was not on at the time, so no one heard the Titanic's calls for help.

In the days and weeks following the Titanic disaster, further inadequacies of wireless regulation surfaced. The 1910 legislation had focused on equipment availability, rather than regulating the use of the airwaves. As such, following the disaster, a handful of instances of radio interference arose, whereby multiple parties communicated on the same radio frequencies at the same time, leading to confusion on the part of the receivers. One notable example of interference led an amateur radio operator to misinterpret "Are all Titanic passengers safe?" as "All Titanic passengers are safe." The operator reported this supposedly good news to the media, which, of course, led to further heartbreak after the truth was reported.

The combined effects of these regulatory shortfalls led to the Radio Act of 1912. The Radio Act was the first legitimate attempt to regulate wireless communications in America. Although it has since been replaced by other legislation, the Radio Act established policies that still govern wireless communication. The Act established once and for all that the government would control the airwaves—no one could transmit on a particular frequency without obtaining a license from the government. A second outcome of the Act was that the government would divide the available frequencies among types of users. For example, certain portions of the spectrum were allocated for military use, while other distinct portions would be available for commercial use. Finally, the Act established the concept of the importance of different types of wireless communications. To illustrate, distress signals were more urgent than commercial signals, which were in turn more important than amateur operators' signals.

By the mid-1920s, commercial radio had become an important means of communication and entertainment. To keep legislation up-to-date, Congress replaced the Radio Act of 1912 with the Radio Act of 1927. The new Act declared that the airwaves belonged to the public, and, as such, access to those airwaves should be granted to individuals and organizations that were most interested in serving the public. Because the spectrum was a public resource, Congress reaffirmed that the government would be responsible for granting licenses to potential transmitters, although those licenses would be free. Between 1912 and 1927, the Secretary of Commerce had been responsible for radio licensing. The 1927 Act, however, determined that this process would be best handled by a special commission. Thus, the Federal Radio Commission (FRC) was created.

Seven years after the last Radio Act, Congress created the most important document ever to regulate communications. The Communications Act of 1934 touched on all aspects of telecommunications at the time, and, in fact, it still remains the center point of all such regulation. (Even other landmark legislation, such as the Telecommunications Act of 1996, only amends the 1934 act, rather than replacing it!) Probably the most important outcome of the 1934 Act was the creation of the Federal Communications Commission (FCC), which replaced the FRC. Five commissioners lead the FCC, which is responsible for overseeing virtually all aspects of communication, from broadcast censorship to radio frequency licensing, in the United States.

One of the most important amendments to the 1934 Act came 62 years later in the Telecommunications Act of 1996. The 1996 Act encompassed a wide range of topics, including cellular, cable, Internet, and other services. One of the main goals of the Act, however, was a push for competition. Where providers had been somewhat limited to specific services, the 1996 Act indicated that a single provider could offer numerous services. Today, we are beginning to see the fruits of the 1996 Act; providers now want to offer a triple play to customers, consisting of voice over Internet protocol (VoIP), cable, and Internet. In some cases, providers might even offer triple play plus one, which includes wireless/cellular service as well.

The regulatory matters discussed above remain the cornerstones of wireless policy. To explore fully any regulatory aspect of even a single telecommunications technology would require a book (or books) of its own. We have kept this discussion brief, but the history of telecom policy is certainly interesting, and we would encourage anyone in the field to investigate it more fully. With an understanding of the history of wireless communication as well as the regulatory factors surrounding wireless communication, however, let us turn our attention to some of the basic technical principles of wireless systems.

Principles of Electromagnetic Waves

The remainder of this book is dedicated to the technical concepts associated with cellular networks. Because wireless communication is at the heart of such networks, it is worthwhile to discuss basic wireless ideas before moving any further.

Wireless systems consist of transmitting and receiving signals through the air (or space) by inducting information onto an electromagnetic wave. To illustrate some of the basic principles of an electromagnetic wave, refer to *Figure 1.1*. Notice that the wave is represented as a function of time and power level. The wave begins at the 0 power level; it moves smoothly up to a maximum level, before returning back to the 0 level; it then moves to a minimum level that is the polar opposite of the maximum level before returning back to 0 again. Upon returning to 0, we say that one cycle of the wave has occurred. The frequency of a wave, measured in Hertz (Hz), is defined as its number of cycles per second. For example, if the second cycle of the wave ended after exactly one second, then the frequency of the wave would be 2 Hz. When discussing radio communications, we are more interested in electromagnetic waves that have frequencies in the thousands-, millions-, or billions-of-Hertz ranges. We refer to these ranges as kilohertz (kHz), Megahertz (MHz), and Gigahertz (GHz), respectively. (Notice that, by custom, the *k* in kHz is not capitalized, although the first letters of MHz and GHz are.)

The electromagnetic spectrum (EMS) consists of all the possible frequencies of electromagnetic waves. For communication purposes, higher frequencies are generally better for faster communication. In addition, we can achieve faster communication speeds by using a wider range of frequencies from the spectrum. Unfortunately, two devices cannot effectively transmit on the same frequencies at the same time, if they are near enough to one another that their signals may interfere with one another. This is the principle problem with wireless communication—the available frequencies in the spec-

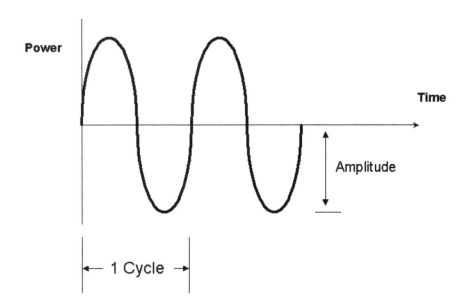

Figure 1.1: Electromagnetic Wave

trum must be allocated in such a way that wireless devices do not interfere with one another.

We will spend a great deal of time in later chapters, discussing these topics in more depth. It is important, however, to keep these basic properties of electromagnetic waves and wireless communication in mind as we move forward.

Conclusion

Whenever one studies an area of telecommunications, such as cellular networks, it is worthwhile to know the history of the relevant technologies. In this chapter, we covered some of the milestones of wireless and telephone communications, including their convergence in the 1940s and the birth of cellular in the 1980s. We also discussed some of the key regulatory factors governing wireless communication, including the Radio Acts of 1912 and 1927, and their successors, the Communications Act of 1934 and the Telecommunications Act of 1996. Finally, we briefly covered some of the key concepts of electromagnetic waves and wireless communication. With this background, we are ready to proceed with a more thorough discussion of cellular networking concepts, beginning in *Chapter 2*, with an overview of cellular systems.

Chapter 2: System Overview

Components of the Cellular Network

To understand the complex structure of the cellular network, it is best examined by looking at its major components individually, then learning how they interact with each other. Terminology for these components can be very confusing as there are various names used by different cellular technologies to define the same unit of the network. Multiple names are incorporated throughout the description to help the reader understand the similarities between the names. There are four major components to all cellular systems:

- The public switched telephone network (PSTN)

- The mobile telephone switching office (MTSO); this can also be referred to as the mobile switching center (MSC)

- The cell site (CS), which is also referred to as the base transceiver station (BTS)

- The mobile subscriber unit (MSU)

These four components are interdependent and are comprised of wireline and wireless technologies. A brief definition of each will help with the under-

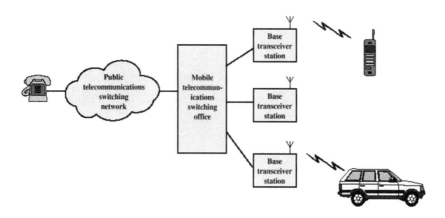

Figure 2.1: *Simple Cellular Network*

standing their relationship. *Figure 2.1* gives a simple overview of the cellular network.

PSTN

The PSTN is comprised of the wires and connections of the outside plant, the local switching or end offices, the trunks connecting the end offices with the tandem and toll switching offices, and the vast array of transmission equipment. The PSTN is the wireline side of the wireless network, responsible for connecting mobile subscribers with fixed wireline users across the world. The switching controls for the network rely on signaling system 7 (SS7) out-of-band signaling network to control circuit set up, monitoring, and call tear down after completion. The PSTN was based on the creation of a direct connection from the sender to the receiver over the network. Each of the sections of the PSTN is an area of expertise unto itself, requiring high levels of knowledge and skill to maintain and configure. It has been established for more than 100 years when Bell first said into his new device, "Watson come here I need you!"

MTSO

The MTSO consists of many subsystems that are responsible for connecting mobile to mobile users, mobile to wireline users, and wireline to mobile users. Housed within the MTSO is an environment that is similar to that of the PSTN. It contains user databases, switching systems, communication trunks between base stations and the PSTN, management terminals, and redundant power systems to maintain the network during electrical failures. The switching systems are responsible for establishing the circuit route in which a call will travel from a subscriber unit to the called/calling party. The database provides for the information related to billing and call tracking. The communication trunks connect the MTSO with either another wireless switching office or the wireline network. Another attribute of most MTSOs is that they maintain some form of redundant communication trunk connection to the base stations either through microwave, satellite, or both. The MTSO is also responsible for cell-to-cell handoff as a mobile subscriber moves through the network. Originally the connectivity from the MTSO to the base stations was over standard T-1 circuits, but the volume based on demand has forced a higher throughput demand to a T-3 (28 T-1s) in major metropolitan areas.

Other functions of the MTSO reside in subsystems within the office. The home location register (HLR) database is located within the MTSO and it is responsible for documenting personal information about the users of the mobile network. The visitor location register (VLR) is also a database located within the MTSO. This database temporarily holds the information for the roaming user within a cell or sector that is being served by the MTSO. Authentication is another function provided by the MTSO. This generally happens when the subscriber powers up their phone and signals the base station (BS) with its electronic serial number (ESN), which is hard coded from the equipment manufacturer for every mobile device. The authentication system will try to verify the mobile user and once verified, it will encrypt all wireless communication between the phone and the network for security purposes. The equipment identity register (EIR) keeps a record of blacklisted mobile phones. This is particularly true and useful regarding stolen mobile phones. The EIR will keep track of and even trace devices on this list.

BS and Base Station Subsystems (BSS)
The BS is generically composed of an antenna, a transmitter and receiver (commonly referred to as a transceiver) for each group of frequencies that

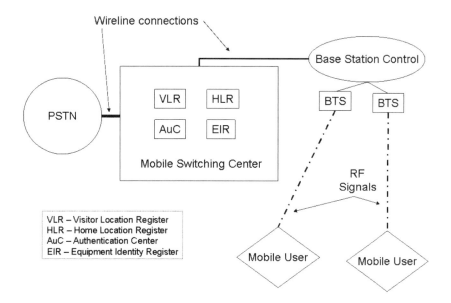

Figure 2.2: Basic Network Architecture

are incorporated for a cell site, and switching equipment to connect calls within the cell and back to the wireline network through the base station controller and the MTSO. The BS is the core of each cell and responsible for sharing frequencies among the users of the area serviced by the antenna. Radio frequency (RF) communication channels are transmitted and received by a base station transceiver (BST). A base station controller (BSC) can be used to control a single or multiple BSs. The BSC is responsible for allocating usable frequencies, handoff control, and signaling mobile users for call notification. To avoid interference across the frequencies used for communication, cells next to each other do not use the same frequencies (defined in greater detail in this chapter). The design of the cellular network is constructed such that the maximum frequency reuse is possible without creating crosstalk between cells using the same frequencies. *Figure 2.2* provides a basic block diagram of the cellular network architecture.

The size (radius) of a cell can vary based on the previously defined parameters (i.e., antenna height, power, and user demand). A base station can have a radius as small as 100 meters and as large as 35 kilometers. A cell site will contain network communication equipment to link back to the MTSO and to process within cell communications. The primary function of the BTS is to manage the delivery of paired frequencies to the mobile user for transmission and reception.

MSU

MSU, which is also known as the mobile equipment (ME), is a device that requires communication links through the radio network of the cellular system. As technology becomes more advanced this statement is designed to cover more than just the conventional cell phone. These devices are mobile within and between networks, communicating to achieve maximum signal strength from the antennas within their physical area. They relay their user information to the network to register the device and establish calling privileges. The ESN is the code embedded in the device's signal that the MTSO uses to validate the user. This number is different that than the called number (i.e., the number someone dials to reach your cell phone), and it can be made to be portable between devices if the information is registered on a subscriber identity module (SIM). A SIM is normally associated with Global System for Mobile Communications (GSM) that is predominant in Europe and used by Cingular and T-Mobile in the United States.

The basic access to the cellular network is similar to that of a landline connection, except RFs are used instead of copper connections. When a mobile user places a call onto the cellular network, the wireless device will send out the radio signals after the user enters the phone number and activates the send button. The data will then be sent to the BS according to which wireless carrier the user is subscribed. After receiving the signal, the base station passes the data onto the MTSO to process. MTSO at this point identifies and records all the information regarding this particular call before sending it to the PSTN to send to either the wireline receiver or across the network to another MTSO to process for delivery to a wireless subscriber. It is necessary for the MTSO to record all information, as this is how the mobile wireless companies keep track of their network traffic. Once this process is fin-

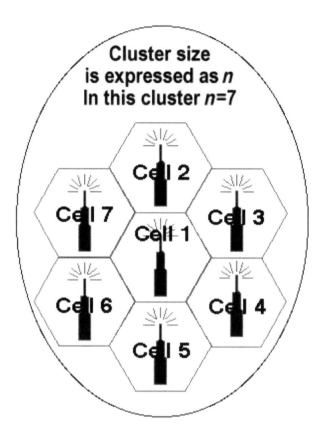

Figure 2.3: Reuse-of-Seven Cluster of Cells

ished, MTSO will forward the data to the PSTN in order to connect the caller to the called party. More detail on the identification and processing of the request for transmission is provided in mobile call connection section.

Cellular Network Organization

Most of the cellular devices that can be found in the market today use a low-power transmitter, which is used to bridge the connection between the mobile device and the cell sites. These low-power transmitters are defined by IEEE standards to transmit at 125 milliwatts (mW) of power with peak output of no greater than one watt (1 W) in 1800GSM systems. Cell sites, which are based on the antenna servicing each site, are theoretically configured in a hexagonal pattern in relation to other antenna geographic cell sites. The actual RF pattern of the signal strength of the antenna does not follow the symmetrical lines of a hexagon. The signal strength can vary on different factors such as physical obstructions (e.g., buildings), geographic variations (e.g., mountains and valleys), and signal interference caused by electromagnetic devices (e.g., high power lines). The design pattern of cell tower placement resembles that of a bee's honeycomb. It is a useful tool in defining how a network can be laid out to service specific areas or regions.

Multiple frequencies for communication are defined by specific licenses granted by the FCC to carriers providing services in geographically defined areas. These frequencies can be reused, with careful engineering to avoid signal interference, within the same area. Each BTS contains a radio transceiver and controller and provides radio communication to the mobile units located in its cell.

The deployment of frequencies in the United States follows a format called the reuse-of-seven configuration. *Figure 2.3* depicts the theoretical design of seven adjacent cells. The next grouping of cells would follow this defined pattern, with adjacent cells not sharing similar radio frequencies. This is done to avoid any cochannel interference from the adjoining transmission tower. Control of power and antenna height defines the reach of the signal and its intrusion into the neighboring cell site.

Two primary types of channels are responsible for communicating with the mobile user: control channels and traffic channels. Control channels provide information between the cellular phone and the cell site on call set up and on sustaining the transmission. Control channels locate mobile devices when there is a call to be delivered. The channel that is being used will then communicate with the MTSO prior to its ringing and assigns which frequency it should be used in order to connect to the receiver. Once the frequency is determined, the control channel will carry the communication information on outgoing calls. The efficiency of the control channel allows for one channel for each antenna site. Traffic channels provide for voice or data communications across the assigned radio spectrum. Two separate channels are provided to a user during transmission: one for transmitting information and one for receiving, which is called full duplex signaling.

Improving Cell Coverage

Cell splitting is the process of subdividing a congested cell into smaller cells, each with its own base station and a corresponding reduction in antenna height and transmitter power. Cell splitting increases the capacity of a cellular system because it increases the number of times that channels are reused. Due to the increasing mobile cellular traffic, mobile wireless providers are subdividing each cell into smaller clusters to keep up with the demand.

In cell sectoring, channels of a particular cell are broken down into sectored groups and are used only within a particular sector. Sectoring reduces the interference, which amounts to an increase in capacity. Sectoring reduces the coverage area of a particular group of channels causing more handoffs. Base stations support sectorization and allow mobile users to be handed off from sector to sector within the same cell without intervention from the MSC or MTSO.

Microcells are yet another reduction in the footprint of the antenna delivery area. Some microcells are characterized by antennas coming down from higher elevations and residing on power poles and light poles at street level. The radiated power from the antenna is also reduced in order to ensure minimal interference with other cells.

Repeaters are devices that extend the RF transmission. Repeaters are bidirectional in nature, and simultaneously send signals to and receive signals from a serving base station. They may send and receive on different frequencies in order to increase the throughput and reduce the delay in transmission. Rural installations rely on the use of repeaters to extend the signal coverage area. They are capable of repeating an entire cellular or personal communication system (PCS) band. Signal delay and the injection of noise in the repeater amplification process may distort the original signal or cause enough lag of the transmission to cause it to be terminated.

Mobile Call Connection

Before a mobile phone user can connect to a called party regardless of where the mobile unit is located, mobile unit initialization is necessary. When the mobile user switches on the unit, the five-digit system identification number (SID) signal will be sent to the nearest base station in order to connect the unit to the provider network. The SID defines the particular carrier that the mobile user is subscribed. Once the connection is established, the mobile user will be able to originate a call by entering the phone number they are trying to reach. When the user presses the "send" button from the mobile unit, the phone number that is entered will get transferred to the base station. After the base station identifies all the information, the paging process will then begin. If the phone number is directed to another mobile unit, the call will be routed to the MTSO to identify the receiving party before it is sent to the base station. The base station sends out a signal to page the mobile unit to notify the unit of an inbound call.

During the paging process, when the connection between the two parties is established, the receiver's phone will ring to signify the incoming call and the receiver can choose to accept or reject the call. If the call is accepted, the channels for transmitting and receiving are connected allowing the call to process. In addition, once the users are connected, MTSO returns the data transferring process of voice or data to the connected network.

MTSO also acts as a control agent with all the collected information while the connection is being established:

- Call blocking prevents users from making certain outgoing calls and permitting incoming calls to connect to the mobile unit if the calls (either incoming or outgoing) contain a certain pattern of digits at a certain time period that is specified.

- Call termination disconnects particular calls that are restricted from connecting to the called party. It can be used when callers are making international calls with a restricted unit.

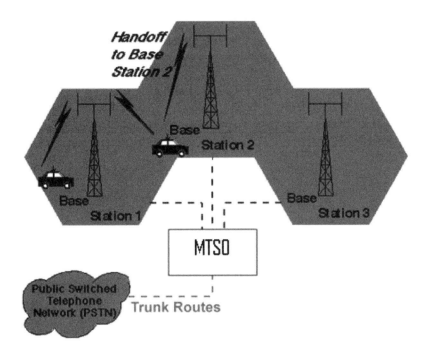

Figure 2.4: *Cell Handoff*

RF Signal Propagation

Signal strength in a cell phone affects the quality of a call. In order to receive or place a call, the cell phone must have a certain level of signal or reception with the base station. The cell phone visually indicates the level of connection by the display of the received antenna power. Because the antenna on each phone is different, the ability to send and receive signals can vary. When the mobile unit cannot connect to the base station, no indication of network or visual power value associated with the antenna is normally displayed. A weak signal does not allow users to receive or place a call, and a signal that is too strong can create interference within the cell to other adjacent frequencies being used.

Signal fading occurs when the distance between the mobile unit and the antenna increases, multiple signals are received causing attenuation or cancellation of the main signal, and where the signal can be obstructed or absorbed by the surrounding environment. Fading causes the signal strength to vary, providing challenges at the receive end of the transmission to accurately decipher the RF signal. Because the user in a mobile environment is not primarily fixed, the change rate of the signal strength is constantly varying and being reassessed by the base station. The challenges from an engineering perspective are great in dealing with a user who could be pedestrian at a very slow rate of change in location to someone driving down the highway at speeds in excess of 60 miles an hour.

Cell Handoffs

Because adjacent cells do not use the same radio channels in cellular systems, a call must either be dropped or transferred from one radio channel to another if a user migrates between the adjacent cells. Because dropping the call is unacceptable, a handoff between cells must occur. A handoff coordinates the connection of a mobile user from one cell site to another based on the type of technology being deployed for the cellular system. Code division multiple access (CDMA) systems work differently than GSM and time division multiple access (TDMA). The handoff from one base station to another is coordinated by the MTSO based on a number of performance metrics that define the quality of service (QoS) associated with cellular communications. The generic qualities, however, are similar.

Call blocking occurs when a call is unsuccessful due to the unavailability of network. The likelihood of this occurring depends on the coverage area of the service provider. The density of users on the particular base station may limit the available channels to be allocated to new users. This is also an issue for a mobile user migrating into a new cell. Call dropping is the likelihood of an ongoing call being terminated. This is usually associated with an ongoing weak power signal or interference on the traffic channels. A call can also be dropped as it is being handed off between cells. If a frequency is not available for a call to be moved from one base station to another, the system has no alternative other than dropping the connection. Base-station planning includes guard frequencies that are held in reserve for transferring calls, however, if traffic is heavy; the likelihood of the available links greatly diminishes.

Signal strength defines the handoff occurrence. If the signal strength from the accessed base station diminishes, the mobile unit needs to connect to a stronger signal to maintain communications. Cellular devices are in constant evaluation mode to determine if there are any other base stations within their receive area that provide stronger signals. The predefined level of acceptance for a signal's strength over a certain period of time provides a safeguard to the call constantly moving back and forth between two similarly strong signals. *Figure 2.4* depicts the handoff of a mobile user from one cell base station to another.

Power Issues

When a cell phone is powered on, it is in constant communication with a base station with which it has associated itself. This constant communication, even though no communications are processing, is a constant drain on the battery life of the cell phone. There is a continuous effort by handset manufacturers to optimize the use of energy and extend the life of a charged battery in a handset.

Power control, which is the receive signal strength of a base station used to define the strongest signal for association of the mobile user, is characterized by two different modes of control. Signal strength must be maintained at engineered levels for all users to help reduce cochannel interference. There are two different types of power control associated with cellular systems. Open-loop power control depends exclusively on the mobile unit without any response from the base station on the power setting. Depending on the signal strength, the mobile will increase its power usage to a higher or lower level. If the signal strength is too strong, the mobile phone will adjust to an appropriate level of its power usage.

Closed-loop power control is controlled by the base station. The base station makes power adjustment decisions and communicates the available information with the mobile phone on the control channel. The sampling rate for the current power value is 800 times per second. Unlike open loop, closed loop depends on the signal strength being controlled from the base station in reference to the power signal received by the mobile phone. The base station sends commands over the control channel based on how well the mobile phone can receive the signals. If the mobile reception is weak, it will tell the mobile to increase its power in order to improve the signal quality.

Traffic Engineering

Traffic engineering in cellular technology is used to maximize the number of available frequencies to the anticipated users within the base station's coverage area. The application of models used for over a century in wireline probability analysis is similar to what is used in wireless; however, radio frequencies of limited numbers are the replacement for trunks in a central-office configuration.

Two types of networks can exist (however, not simultaneously) for cellular systems: blocking and nonblocking. Blocking systems block calls when the maximum capacity is reached or, if there are assigned guard frequencies, calls are blocked when there are a minimum number of frequencies left available for processing handoffs from another base station. Calls can also be blocked based on the trunk capacity of the wireline system feeding the base station. Nonblocking systems allow mobile users to place or receive calls without experiencing denial of access to any frequencies, features, or functions on the network. This is accomplished by ensuring the number of concurrent users does not exceed the available channels. This is accomplished by limiting the number of possible subscribers to the system and having excess capacity in regards to available communication channels. When the simultaneous users are over the capacity the network can accommodate, however, then calls will be denied. When this occurs, the system is blocking.

The network is designed to a specific level of performance based on parameters such as calls blocked, number of calls dropped, service busy, and dropped handoffs. The QoS that customers demand drives the placement of additional cells or the subsetting of existing cells to create more frequency

SSD (Shared Secret Data) :

- 128 bit secret number that is semi permanent

•Input for authentication calculations in the phone and AC

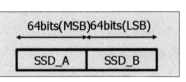

•Can be modified over the network

Figure 2.5: Shared Secret Data

opportunity for the subscriber population. The grade of service is based on the probability of a call being blocked. The lower the probability, the better the QoS is considered on the network.

Traffic intensity is used to measure the average occupancy of a facility during a period of time. With the traffic intensity, engineers can decide the amount of resources and equipment that are needed for a higher QoS. An important aspect of traffic is that it is considered as the connection between the two parties regardless the amount of information that is being transmitted, either voice or data.

There are several ways to handle blocked calls other than simply deny access to the network. One of the methods used is called "lost calls delayed." When the intensity limit is reached, rather than giving callers a busy tone, callers may experience a delay or a long pause before the call can be connected. It basically puts the caller's request for services in a queue and connects the caller as soon as a channel becomes available. When this happens, callers usually will experience a long pause before the call can be connected. Another way to handle blocked calls is called "lost calls cleared." In this technique, a call will be held until it reaches the receiver. Usually a message such as "Retry?" will come up on the mobile phone for callers to decide whether they want to continue the repeated attempts to connect the call.

Security

In earlier generations of cellular technologies (e.g., advanced mobile phone system [AMPS]), the information concerning the user's electronic identity was broadcast openly, which made the technology susceptible to fraudulent practices such as the capturing of the ESN to be transferred to another chipset on a different mobile handset. Second-generation mobile technologies have provided methods for securing information as it is transferred. All second-generation forms of cellular service is digital in format, allowing the systems to encrypt the data as it is transferred. Encryption is an effective way to achieve security. To read an encrypted file, you must have access to a secret key or password that enables you to decrypt it. Unencrypted data is called plain text; encrypted data is referred to as cipher text.

Encryption can perform the following:

- Reversibly transforms a bit stream into another so that any reasonable duplication of the original bit stream is unavailable to a receiving terminal without knowledge of appropriate keys

- Effectively prevents fraud

- Prevents eavesdropping on cellular phones

Authentication encompasses the following:

- Process for confirming the identity of the mobile station

- Prevention of fraudulent use of the network by mobiles programmed with counterfeit MIN and ESN

- Cellular authentication and voice encryption (CAVE), which is an algorithm for authentication and key generation

When each mobile device tries to access the network, it has to go through an authentication process to confirm that the identity of the calling mobile station is valid. The purpose for this authentication process is to prevent fraud with the valid mobile identification number (MIN) and the ESN. The MIN and ESN are burned into the chipset associated with each mobile device, providing a unique identifier to the system. CAVE is an algorithm that exists on all second-generation cellular phones and is used to provide secure access to the wireless network. A secure key from the user is matched with the values it transmits to the network, where the information is examined with the authentication key. If the values are the same, the user is allowed access to the network. The following are true for CAVE:

- The CAVE algorithm is used for authentication in North America

- CAVE is comprised of shared secret data (SSD) in the cellular unit and at the authentication center

- CAVE uses authentication key (A-key) and ESN to produce a secure server network (SSN)

Descriptors of A-key are as follows:

- Comprised of 64–bit secret number that is permanent

- Used to generate the SSD

- Stored in the mobile (it is never sent over the air)

- Programmed into the phone during subscription

The A-key is a secret value that is unique to each individual cellular phone. It is registered with the service provider and stored in the phone and authentication center (AC). The A-key is programmed into the phone by the manufacturer. It can also be entered manually by the user, from the phone's menu, or by a special terminal at the point of sale.

The phone and the AC must have the same A-key to produce the same calculations. A-keys are never sent over the air interface or the SS7 network. The user should not be able to display the A-key on the phone. The primary function of the A-key is to be used as a parameter to calculate the SSD.

SSD includes the following (see *Figure 2.5*):

- A 128–bit secret number that is semipermanent

- Input for authentication calculations in the phone and AC

- Ability to be modified over the network

The SSD is used as an input for authentication calculations in the phone and AC and is stored in both places. The SSD can be modified over the network. It can be sent over the SS7 network, but will not be sent over the air interface. The SSD is updated when a phone makes its first system access and periodically thereafter. When the SSD is calculated, it results in two separate values: SSD–A and SSD–B. SSD–A is used for authentication. SSD–B is used for encryption and voice privacy. Because this SSD is a 128–bit secret number, SSD–A and SSD–B both have 64 bits each.

Conclusion

The simplicity with which we use our cell phones today hides a very complex system of access, RF signal propagation, and security. The quick migration of cellular technologies to their present state has evolved in this country in just over 20 years. Future configurations of the network for wireless technologies will be based on the fundamentals used to establish the existing network. Next-generation technologies, such as end-to-end IP packetized voice and data in the network, will still require the user to access the network, receive RF channels, and be authenticated with the provider

Chapter 3: Radio Signals

To appreciate any telecommunications system fully, it is important to understand how information signals are transmitted within that system. In wireless networks, information is transmitted by way of radio frequencies. But, how does this happen? We have already examined important properties of radio frequencies. In this chapter, we will discuss various methods of transferring information on radio signals.

Analog vs. Digital

The term "analog" comes from the idea that a transmitted signal is analogous to the original signal. A classic example of an analog technology is the last mile of the plain old telephone service/system (POTS). The average person's speaking voice will fluctuate between a wide number of frequencies between 0 and 4000 Hz, as well as a wide variety of volume levels. An analog telephone set will represent each of these infinite possible combinations of frequencies and volume levels with a similar electrical signal that is propagated to the provider.

In a world of perfect communication media, analog technology would be fine; unfortunately, we do not live in such a world. Regardless of the quality of the media that is used, noise will always be present in a communication link. Noise can be considered any unwanted electromagnetic interference that is introduced into a communication circuit. In an analog system, it is impossible to know for certain which portion of a received signal is real information and which is noise. To compound the problem, in analog systems that require amplifiers at regular intervals (such as POTS), not only is the original signal amplified, but the noise is amplified along with it. Therefore, if more than a few amplifiers exist between the transmitter and the receiver, it becomes impossible to distinguish real information from noise.

To combat the problem of noise in analog communication, digital technologies were developed. With digital systems, the original signal is converted to one of a finite set of values before being transmitted. Because these values are discrete, it is possible for a receiver to accurately determine what value was transmitted. For example, let us suppose that a digital device may transmit one of two values on a communication circuit: +3 volts or 0 volts. If the receiver detects +2.38751 volts, it determines that this number is closer to +3 than it is to 0; as such, the receiver treats this input as +3 volts. In an analog system, the receiver would have to assume that the transmitter intended +2.38751 volts.

In the case of voice communication, there is no such thing as an all-digital network, because the human voice has an infinite range of possible frequency and loudness combinations. Even with digital wireless phones or VoIP technologies, there is an analog-to-digital conversion process that must occur within the digital telephone. In the case of data communication, however, networks can be all digital (and typically work best when no digital-to-analog conversion is required). Because our focus is currently on voice communications, this is an important point to keep in mind.

The examples above have focused on wired communication systems, but the same is also true for wireless networks. Noise is a much greater concern for wireless than it is for wireline communication, so digital techniques are becoming ubiquitous for cellular networks. Original systems, however, were actually analog by nature. For example, the AMPS was a first-generation cellular technology that utilized analog signal transmission.

Modulation Techniques

The concept of modulation is imperative to understand how communication can occur via radio frequencies. Essentially, modulation is the process of altering some aspect of a carrier frequency (or carrier signal), such that communication may occur. So then, what is a carrier frequency? Although you may have never heard the term before, you are likely familiar with the idea. For instance, consider your favorite radio station—what channel is it on? 106.1? 103.3? 92.9? These numbers refer to the carrier frequency of the radio station, in Megahertz. So, when you tune in to 106.1 to listen to your favorite classic rock tunes, your radio receiver locks onto all signals coming in at (or near) the carrier frequency of 106.1 MHz. Because any given carrier frequency is just a sine wave, it must be altered in some manner so that information may be put onto it; this process is called modulation. In our example, your radio locks onto the frequency 106.1 MHz, and looks for variations in that frequency to demodulate the signal and output the intended signal.

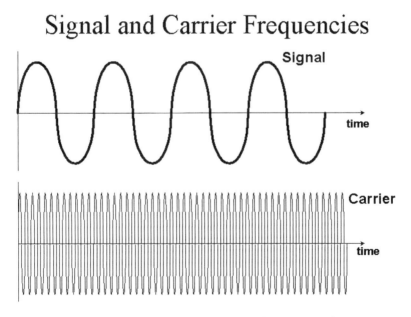

Figure 3.1: Comparison of Signal and Carrier Waveforms

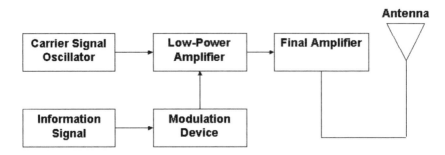

Figure 3.2: AM Transmission System Block Diagram

There are a number of modulation techniques that can be used to transmit information across a radio medium, and we will discuss the most common of these techniques. One commonality between techniques, however, is that the carrier frequency will be much higher than the information signal (sometimes called the baseband signal). For example, compare the baseband signal and carrier signal in *Figure 3.1*. This allows receivers to accurately derive the baseband signal from the received signal.

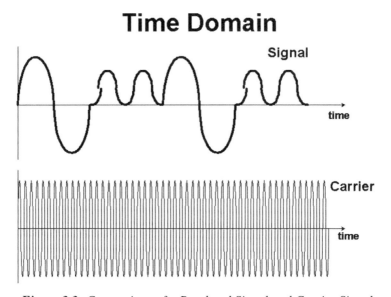

Figure 3.3: Comparison of a Baseband Signal and Carrier Signal

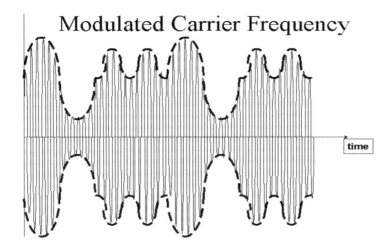

Figure 3.4: AM of Baseband and Carrier Signal

Three common modulation techniques are worth further discussion: amplitude modulation, frequency modulation, and phase Modulation.

Amplitude Modulation

Amplitude modulation (AM) is the most commonly used radio frequency modulation. In this type of modulation, the amplitude of the carrier wave is varied in direct proportion to the baseband signal. The information signal is

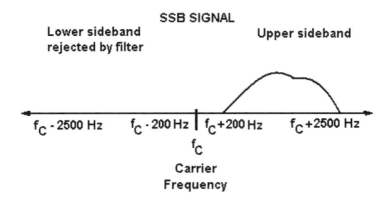

Figure 3.5: Single Sideband Signal

mixed with the carrier signal in a frequency mixer, as illustrated in *Figure 3.2*. The output of the mixer thus contains the modulating signal whose frequency is equal to that of the carrier signal but with peaks and troughs that vary in proportion to the strength of the information signal. The modulated signal is then amplified and fed to an antenna for transmission. Although it is more susceptible to noise than other techniques, AM was the first modulation method used in commercial radio broadcasting and continues to be prevalent in modern wireless applications.

Figures 3.3 and *3.4* represent the AM process, as a combination of a baseband signal and a carrier signal. *Figure 3.3* shows the carrier wave along with the information signal. These two signals are fed into the frequency mixer, and the output thus obtained is illustrated in *Figure 3.4*. Thus, from the figures observe that the amplitude of the carrier signal changes in accordance with the information signal.

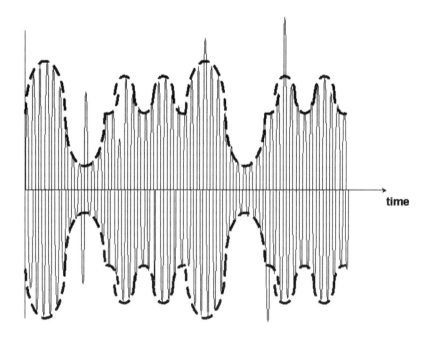

Figure 3.6: *Noise in an AM Signal*

The waveform in *Figure 3.4* represents the signal that would be sent from the transmitter to the receiver. Notice that the original information signal is traced on the amplitude of the carrier frequency. Also, observe the mirror image of the baseband signal on the negative side of the carrier signal. This envelope of information can be reduced to a single side of information by filtering out the lower half of the composite signal represented in *Figure 3.4*. By utilizing a single sideband (SSB) carrier, AM systems can use lower power transmitters, allowing for a more efficient system. *Figure 3.5* represents the resultant SSB signal.

AM makes very efficient use of the scarce radio spectrum by limiting transmission to discrete frequencies (as compared to frequency modulation, which will be discussed next). Unfortunately, though, amplitude modulation is highly susceptible to noise, while some other techniques are not. Any high-power electromagnetic interference that is present in the atmosphere between the transmitter and receiver can add to the AM signal. This added power raises the amplitude of the signal beyond a recognizable range and distorts the baseband signal that is recovered. This is the reason why when

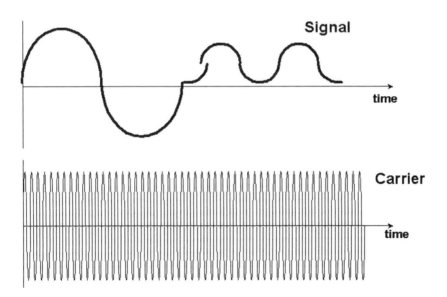

Figure 3.7: Comparison of a Baseband Signal and Carrier Signal

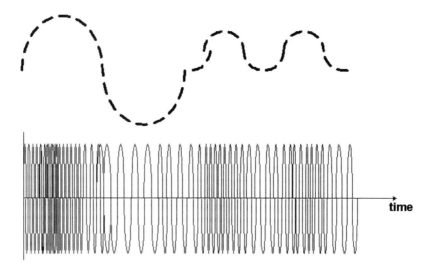

Figure 3.8: Frequency Modulation of Information and Carrier Signals

listening to an AM radio station during electrical storms, we hear static on our radio receivers. *Figure 3.6* represents noise added to an AM signal. Note the spikes of power above and below the signal envelope. The receiver detects these power surges as noise.

The efficiency and simplicity of an AM transmission made it possible to implement in stationary and mobile configurations. Though surpassed by other modulation techniques for quality of sound applications, it is still a viable technology for delivering information. It is also worthy to note that although the figures above represented an analog baseband signal, it is common to transmit digital signals via AM. For example, in a simple system, we might transmit a discrete amplitude for a given period of time to represent a 1 and another discrete amplitude to represent a 0. More advanced systems, such as quadrature amplitude modulation (QAM) use multiple discrete amplitudes, along with phase shifts, to represent multiple bits at a time.

Frequency Modulation

Another common modulation technique is known as frequency modulation (FM). As the name suggests, FM is accomplished by varying the frequency of a carrier signal to transmit information. As with AM, a baseband signal and carrier signal are combined in a mixer, but the characteristics of the resultant signal are different from those of an amplitude-modulated signal. With FM, the amplitude of the transmitted signal should be uniform. As the baseband signal's power (amplitude) increases, however, the carrier signal's frequency increases. Conversely, when the baseband signal loses power, the frequency of the carrier in turn decreases. Thus, by its very nature, FM transmission requires a range of frequencies, as opposed to AM, which requires only a single frequency for transmission of a signal. As such, FM requires a wider bandwidth than does AM. FM is illustrated in *Figures 3.7* and *3.8*.

Perhaps the greatest advantage that FM offers over AM is its relative resistance to noise. As previously discussed, noise in AM results from spikes in power (amplitude) in the received signal. With FM technology, however, amplitude is not examined at the receiver—only the frequency is of consequence. Therefore, FM receivers can filter momentary power spikes more easily than AM receivers.

As with AM, FM may be used to transmit either analog or digital information signals. In the case of digital transmission, FM is typically referred to as frequency shift keying (FSK).

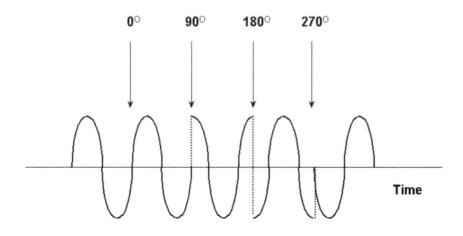

Figure 3.9: Phase Modulation

Phase Modulation

Phase modulation (PM) is achieved by varying the instantaneous phase of the carrier wave, as a result of variations in the baseband signal. Unlike AM and FM, PM is most practical with digital signal transmission, in which case, it is called phase shift keying (PSK). To further understand this, let us consider the carrier signal as a sine wave. Because a sine wave is defined by degrees ranging from 0 to 360, we can think of the sine wave as a complete circle; the cycle of a wave is defined by how often it repeats over a specific period of time. If we were to imagine a sine wave and trace its progress throughout the 360 degrees, we would see the following: it would begin at the intersection of the x-axis at 0 degrees; it would then travel up to a maximum point of 90 degrees; it would then return down to the x-axis at 180 degrees; it would then progress down to a maximum point of 270 degrees; and, it would then return back up to the x-axis at 360 degrees.

Because the 360 degree mark could be considered the beginning of another repetition of the wave, there are essentially four landmark levels in a single instance of a sine wave: 0 degrees, 90 degrees, 180 degrees, and 270 degrees. With PM, a transmitter will alter the phase of the transmitted signal to signify a change in the baseband signal. For example, we might represent a binary 00 by changing the phase of the signal to 0 degrees; 01 might be 90 degrees; 10 could be 180 degrees; and 11 could be 270 degrees. As such, a baseband signal of 10 00 01 would correspond to a sine wave that began at 180 degrees and continued for a predetermined length of time; then instantaneously moved to 0 degrees and continued for a predetermined length of time; and finally moved to 90 degrees and continued for a predetermined length of time. For a graphic representation of PM, refer to *Figure 3.9*.

Of course, given transmitters and receivers of adequate complexity and ability, other degree levels could be used—however, it is important to ensure that receivers will be able to differentiate between the intended phases. For instance it is easy enough for a receiver to know the difference between 90 degrees and 0 degrees; it would be less easy for a receiver to differentiate between 45 degrees and 46 degrees. This is why PM is well suited for digital transmission.

Multiplexing

In general, it is desirable to allow as many channels of communication as possible over a single physical circuit. This is true whether that physical circuit is a 4-wire copper cable in a telco circuit, a coax drop from your cable provider's plant to your home, or the air that serves as the transmission medium in a wireless application. Multiplexing is the term used for any technique that allows multiple communication channels to coexist simultaneously on a single medium. Two forms of multiplexing are widely used in telecommunications: frequency division multiplexing (FDM) and time division multiplexing (TDM).

FDM

FDM techniques allow multiple simultaneous communication channels to exist on a given medium by separating them into different frequencies. For instance, one channel might operate at 100 MHz, the next might operate at 200 MHz, and so on. As such, FDM is a natural technique to use in wireless applications. In fact, frequency division is so critical in wireless communications that the federal government must grant a would-be wireless user the right to use a particular frequency before he or she may use that frequency. Note that FDM is typically referred to as frequency division multiple access (FDMA) when used in conjunction with wireless or cellular networks.

FDM techniques are also used in some wireline communication technologies as well. The most notable of these is cable television (CATV). With CATV, the cable that plugs into a television set essentially mimics the antenna on traditional television sets. Of course, this cable does not just pull signals off the air, however—it has a direct connection to equipment at the cable office head end that places television feeds onto different frequencies on the cable itself.

TDM

TDM is by far the most common form of multiplexing outside of the wireless communication arena. TDM is most useful when communication channels are not divided into different frequencies—instead baseband signals are transmitted directly on a medium. In this case, each communication channel is given a regularly occurring timeslot, during which it has complete control of the medium. This is the foundation on which many familiar telecommunication technologies are based. For instance, T1, T3, E1, and integrated services digital network (ISDN) primary rate interface (PRI) are all TDM technologies, based on a 64 kbps building block, known as digital signal level 0 (DS0).

Despite the fact that TDM techniques are most commonly used in telecommunication backbone circuits, they may be applied to wireless applications, in the form of TDMA. In TDMA environments, a wireless user (such as a cellular phone) is able to detect any unused timeslot on a given frequency within a cell. Upon doing so, it will begin to transmit during this empty timeslot—in this way, the same frequency is able to carry multiple communication channels.

Conclusion

In this chapter, we added to some of the introductory material in earlier chapters to better understand exactly how information is conveyed using radio signals. Various modulation techniques exist for transmitting analog and digital information on carrier frequencies within the electromagnetic spectrum. As their names suggest, amplitude modulation, frequency modulation, and phase modulation allow information transfer, based on alterations to some key component of the carrier wave. We also discussed the two primary methods for multiplexing information streams together to make the most efficient use of the transmission channels that are available to us.

Chapter 4: Generations of Cellular Technologies

First-Generation Analog

In early mobile technology, one central tower was placed that could broadcast for an extended distance. All mobile devices used the same channel. Anyone within the range of the tower would connect and channels might not be reused. Eventually mobile communications were divided into a series of cells to eliminate much of this interference and to allow for greater use of the available spectrum.

Developed by Bell Labs in the 1970s, the first-generation analog cellular technology AMPS became the standard in North America and in other parts of the world during the 1980s. Other variations of this technology were deployed in other parts of the world. Nordic Mobile Telephone (NMT) was deployed in Sweden and Norway. Similarly Total Access Communications Systems (TACS) was deployed in Europe and some parts of Asia.

These early systems had a number of drawbacks. The analog systems were not secure in the sense that second- and third-generation cell phones are, because they were analog and had no means of encrypting the call. They also were not efficient in the use of spectrum, requiring large portions of frequency bandwidth that could not be adequately shared between users within a given cell.

AMPS is not capable of transmitting data as we know it today. As such, text messaging and caller ID were not available until digital technologies emerged.

AMPS Operation

AMPS systems were worked very similar to landline operations in how they established, transmitted, and disconnected a call. The MTSO (see *Chapter 2*) was responsible for maintaining the database to authorize access to the network, handoffs between BSs, billing services within and across networks, and monitor roaming between service providers.

Frequency division multiple access (FDMA) was the analog format used to deliver cellular services. The allocated 416 channels of 30 kHz bandwidth were used as transmit, receive, or control channels. FDMA is a multiple access method in which users are assigned specific frequency bands. The user has sole right of using the frequency band for the entire call duration. The spectrum from 824 MHz to 894 MHz was used for the RF transmission. The one advantage of this spectrum was that it had the ability to penetrate buildings for better transmission than the currently used 1.8 and 1.9 GHz

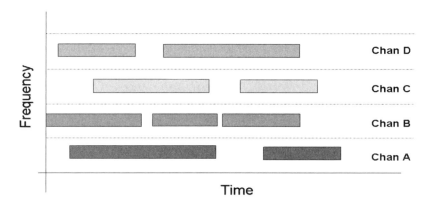

Figure 4.1: FDMA

band in digital cellular services. *Figure 4.1* shows the relationship between users, frequencies, and time in a FDMA network.

A wireless transmission made using FDMA would allow one voice call per channel. Callers would get their own distinct radio channel that would not be used by any other person for the duration of their call. A FDMA channels are 30 KHz wide and are classified as receive channels and send channels. This is the basis of AMPS communication.

Of the ways to share radio spectrum today, FDMA is currently the oldest. While it is a key component in analog mobile communications, its scope goes further into the digital world. That is, FDMA will also be used in combination with CDMA and TDMA.

AMPS was the original mobile-phone technology employed in the United States when it debuted in the early 1980s. This was purely an analog signal and had been used for many years prior to the rise of newer digital technologies. AMPS produced good voice quality, but it was inefficient and not secure. The FCC has mandated that AMPS be phased out by 2007.

Narrowband AMPS (NAMPS) is an enhanced variation of the AMPS mobile-phone technology. This progression provides additional voice-channel capacity than its older counterpart and supports dual-band communication. Phones that harness the NAMPS technology are also compatible with the older standard.

The MTSO is linked to the base stations and the carrier central office by either microwave or landline. Every cell has equipment for radio telephone control and also a control channel and voice channels. The important part of this is the control channel because of the data that it transmits. Also, the control channel is directly involved in communication with each individual cell phone. The data from the control channel is significant because it can either allow the phones to make an outbound call or to notify the cell phones of an incoming call. This control channel is used by the MTSO to assign a voice channel to a call using a 25 MHz frequency.

Limitations for AMPS

In mobile communication there is an information signal that goes between the phone and the network provider. AMPS did not use this signal efficiently at all. In other words, in a given bandwidth with the analog system, the bandwidth was not used to its fullest potential, limiting one voice user as traffic across each channel that was available. The analog communication could not be encrypted, unlike the digital (1s and 0s or binary-formatted information) transmissions of second- and third-generation networks. AMPS also could not provide for high levels of error detection or correction because of the analog delivery. With the migration to second-generation systems, these issues of inadequate network utilization were addressed.

Differences Between First- and Second-Generation Systems

While many mobile devices claim that they are digital, they still use the analog towers when communicating. The phone call is made using the same process as an analog call, but there is a key difference. Calls made on a digital phone are converted into a digital format and then transmitted over the same frequencies that make analog calls. The digital phones take a single analog voice channel and maximize the use of the bandwidth allocated to allow additional users within the same frequency. In second-generation systems, one notable exception exists: PCS. This technology is pure digital. PCS operates on a different frequency than analog systems and generally uses different antennas to transmit as well. In this pure digital environment, the signal is encrypted for security and error correcting occurs to the signal to ensure a clear call from one caller to another. Second-generation mobile cellular systems, also commonly referred to as 2G, employ new wireless networks such as GSM, TDMA, and CDMA. In addition to voice calls, data transmissions were made possible. The ability to send facsimile, text messages, and basic Internet access were made available through network additions within these 2G structures. The primary differences between first- and second-generation systems are as follows:

- Digital communications: The traffic channel, which is the path in which your voice travels, has become pure digital; this means that

Time Division Multiplexing

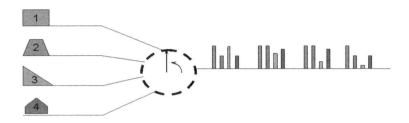

Rotation Analogy

Figure 4.2: In TDM the Rotation Analogy Defines Specific Time Slots on Points of the Wheel Shared by Various Users (Geometric Shapes 1 through 4) at a Predetermined and Fixed Time Interval

your analog voice conversation has been made into binary 1s and 0s that represent your conversation and are transmitted through the air to a receive point that can unscramble the digital transmission to convert it back to sound like your original voice. The transmitted signal, the actual RF propagation, is still an analog process, however the information carried on the radio frequency is digitized. Wireline traffic has been doing this since the T-1 was introduced for backhaul and intraoffice connections in the 1960s. Providing this conversion to wireless traffic has a natural progression of improving the signal propagation and the quality of the signal.

- Increased utilization of spectrum: With analog systems, there was limited access for users based on the FDMA protocol for channel access. In 2G systems, the utilization of the available spectrum was greatly increased because of TDMA and CDMA. Just like landline systems conversion in the 1960s, the conversion from an all-analog network to one that was shared with TDM (see *Figure 4.2*) designs for central office and private automatic branch exchanges, the network utilization was greatly increased.

- Redundant signaling: Transmitted digital signals have forward error correction (FEC). The transmitting end adds coded information to the front end of the packet to allow the replication of the data packet if there is corruption (i.e., errors) of the signal in transmission. Analog systems could not provide this feature; it could not request corrections of errors in transmission because of the synchronous nature of analog voice communications.

- Security: Inherent with transmission of digital information is the ability to secure the transmission with encryption. Analog signal transmissions were broadcast in the clear, very similar to what a radio station would do. If someone had the receiver, they could easily monitor a first-generation conversation.

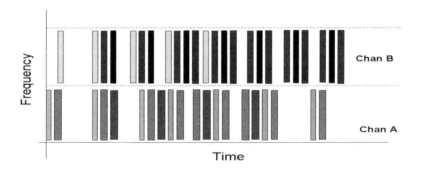

TDMA

Time Division Multiple Access

Figure 4.3: With TDMA Each User Is Represented by a Different Color Within the Frequency Band; Note the Multiple Users in Each Channel in TDMA Versus the Single Users in FDMA

Second-Generation Systems

GSM originated in Europe in the early 1980s. In 1982, a conference of tele-com administrators from 26 European countries was held. A standard means of cellular technology emerged from that conference that would later be deployed among all European member states. The consensus from the very beginning was that the future of cellular communications in Europe would be digital. Phones could be sold to consumers less expensively, and the new digital handsets would have capabilities that their analog predecessors could never support. Deployment of this new technology occurred on July 1, 1991. Over the next decade, it became the predominant method of cellular service delivery in the world.

The Japanese government-owned telecommunications agency, Ministry of Posts and Telecommunications, developed in 1989 a digital platform for their next-generation mobile communications known as personal digital cellular (PDC). It is an efficient TDMA–based access, second-generation TDMA–based service. It does not have nearly the subscriber base of its larger counterparts yet PDC amounts to roughly 10 percent of cellular usage worldwide. PDC uses the 810–826 MHz coupled with the 940–956 MHz

CDMA

Code Division Multiple Access

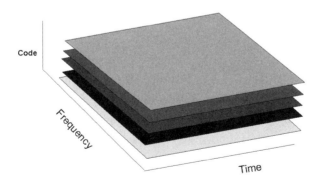

Figure 4.4: *Each Mobile User Is Represented by a Different Layer (and Color) Within the Same Frequency and Time Space; Separated Only by the Data for Each User Being Coded in a Different Way*

spectrum and the 1429–1453 MHz together with the 1477–1501 MHz spectrum allocations for delivery. It uses 25 kHz spacing which helps reduce interference and allows for a frequency reuse of four instead of seven.

TDMA was the TDM equivalent of the wireline migration to digital services (see *Figure 4.3*). TDMA uses the same bandwidth allocations from the FDMA environment; however, it cuts the frequency into time slots shared by multiple users. TDMA transmits in bursts, rather than a continuous signal unlike AMPS. Receiver units must be synchronized for each burst of data. Users must be separated by frequency guard bands which cause more overhead than FDMA–based systems, but it is more efficient because of the time-slot process.

With TDMA, several calls are placed onto one frequency by time slot into a frame. This is similar to the TDM process described previously. Each time slot is allocated to a specific user. The timing of the receive signal is critical in TDMA because each slot represents a different conversation within the frame. AMPS systems have a frequency width of 30 KHz. TDMA is identical in this regard. The guard bands described in previous paragraphs facilitate the separation of the incoming data stream, which provides a buffer for the receive antenna to demultiplex the signal with a reduced error rate and minimize the intersymbol interference (ISI).

Narrowband CDMA uses a spreading process across the allocated spectrum to encode users' data on the same frequency at the same time. *Figure 4.4* provides a graphical presentation of the shared time and frequency approach. This method was originally designed and implemented for military applications that required a high level of security in the transmission and provided a signal that was resilient to jamming or interference attempts from unfriendly sources. Second-generation CDMA operates across a 1.25 MHz section of spectrum. Each user has a unique and separate code, created by using Walsh codes for a random processing, while communicating over the network. Because of its spectral efficiency, CDMA is the platform for third-generation mobile wireless systems. Second-generation CDMA has a number of factors that has separated it from other forms of mobile communications. They are as follows:

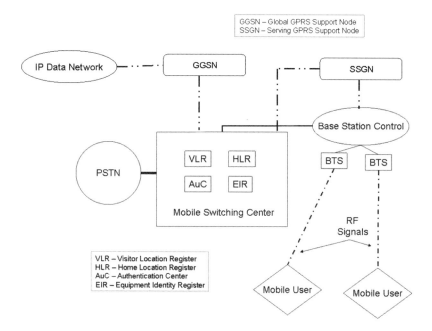

Figure 4.5: *GPRS Overlay in GSM*

- Power control: There is a mechanism used to balance the power of all users within a sector such that communications from distant users are not compromised by users who might have a higher power output closer to the base station antenna.

- Security: With the ability to spread the broadcast information across a wide segment of spectrum in an encrypted signal, CDMA is inherently secure.

- Frequency diversity: Because the information is spread across a wide portion of spectrum, interference or noise has less of an impact on the signal. Noise normally affects ranges of frequencies, not an entire bandwidth of transmission. By allowing the information to be replicated across the 1.25 MHz bandwidth the signal can have portions corrupted but still retain the ability to retrieve the original information on the receive end of the signal.

- Multipath resistance: The complexity of the CDMA signal and its ability to sort through more than one received signal help it to overcome issues that attenuated analog systems. A transmitted signal normally does not have a line of sight (LOS) direction to the mobile user, which has the transmitted signal reflecting, refracting, and scattering through and around man-made and natural obstructions. The delay caused by these obstructions creates multipath signal propagation.

2.5G Cellular Networks

General packet radio service (GPRS) was developed as an overlay of the GSM voice network to provide connectivity for data transmissions. Voice traffic is characterized by numerous blank spots, i.e., times where no conversation is occurring and where the network is not being used. GPRS takes advantage of these unused portions of the air spectrum by multiplexing data transmissions in the available time slots.

GPRS can work with different data protocols, especially those within transmission control protocol (TCP/IP) environment, making it a widely applicable data service. The speed of the connection however, is constrained by the coding schemes used to compress and decompress voice services. Throughput roaming speeds are estimated at 114 kbps. This is a respectable data flow considering the application, but when we consider that the minimum wireline desktop computer in an office environment operates at 10 Mbps, an order of magnitude faster, the GPRS speed is seen as a stopgap measure to emerging wireless data services.

GPRS is termed *2.5 G* as a migration point of adding to the existing second-generation network without fully upgrading the mobile network to 3G technologies. GPRS does not require any special connection or setup control for point-to-point service; it can operate within the existing architecture. The implementation of GPRS into an existing GSM network requires a device to control the data packet injection into the available time slots (packet data unit [PDU]). Two additional devices are brought into the network. They are the serving GPRS support node (SSGN) and the gateway GPRS support node

(GGSN). These devices can be considered as the points where router function would occur in a wireline network. The home location register (HLR), the database that tracks users in the wireless network, is enhanced for GPRS and is responsible for maintaining routing and transfer information for users. *Figure 4.5* provides a block diagram of the additional components in a GPRS overlay in the GSM circuit switched environment. Note the minimal change in the GSM architecture.

The key user features of GPRS are as follows:

• Speed of access: GPRS can achieve theoretical speeds (114 kbps) of up to three times as fast as the data transmission speeds possible over today's fixed telecommunications networks and 10 times as fast as current circuit switched data services on GSM networks. By allowing information to be transmitted more quickly, immediately, and efficiently across the mobile network, GPRS may well be a relatively less costly mobile data service compared to short message service (SMS) and circuit switched data.

• New and better applications: GPRS facilitates several new applications that have not previously been available over GSM networks as a result of the limitations in speed of circuit switched data (9.6 kbps) and message length of the short message service (160 characters). Internet applications ranging from Web browsing to chat can be enabled over the mobile network. Other new applications for GPRS include file transfer and home automation: the ability to access and control in-house appliances and machines remotely.

• GPRS–enabled mobile devices: To avail the features of GPRS technology, the user must possess a GPRS–enabled mobile device. This in itself is an up-grade from the ordinary mobile devices used for the prior GSM technology. A special subscription should be sought from the service provider. Because GPRS involves the usage of Internet, each mobile device is designated an address similar to an IP address.

The key network features are as follows:

- Packet switching: GPRS technology involves the overlaying of a packet-based interface over the already existing GSM network. This way the user has an option to choose the service. In circuit switching a data connection establishes a circuit and reserves the full bandwidth of that circuit during the lifetime of the connection. GPRS is packet-switched, which means that multiple users share the same transmission channel, only transmitting when they have data to send. The information to be transmitted is divided into packets that are reassembled at the receiving end.

- Spectrum efficiency: GPRS services are used specifically when a user is sending or receiving data. The available radio source can be shared at the same time by different users rather than dedicating it to a particular user over a specified time slot. This efficient use of scarce radio resources means that large numbers of GPRS users can potentially share the same bandwidth and be served from a single cell, the area covered by a base station. The actual number of users supported depends on the application being used and how much data is being transferred.

Time Division Duplexing

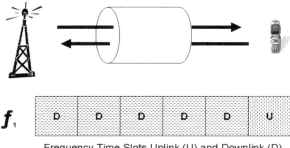

Frequency Time Slots Uplink (U) and Downlink (D)
On one frequency (f_1)

Figure 4.6: *TDD Frequency and Time Slot Allocations*

Frequency Division Duplexing

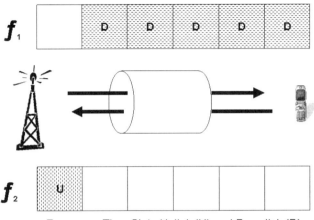

Frequency Time Slots Uplink (U) and Downlink (D)
On frequency one (f_1) and two (f_2)

Figure 4.7: *FDD Frequencies and Time Slot Allocations*

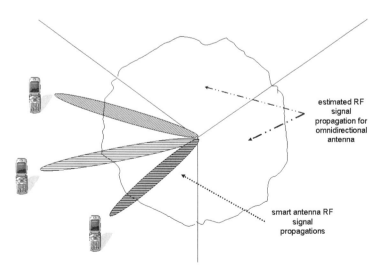

Figure 4.8: *Smart Antenna Signal Propagation*

- Support for TDMA and GSM: GPRS is not valid only over a GSM network. It can also be deployed over the IS–136 (TDMA) standard, which is popular in North and South America. This follows an agreement to follow the same evolution path towards third generation mobile phone networks concluded in early 1999 by the industry associations that support these two network types.

Enhanced Data Rates for GSM Evolution (EDGE)

EDGE, a 2.5G technology, is an improvement on the GPRS packet-switched delivery over TDMA–based networks. EDGE was designed to work with the existing GPRS infrastructure wherever possible, providing improvements on link quality and reliability of data throughput. EDGE has enabled improvements over the radio link by improving the signal to interference condition of each frequency used and implementing a new modulation scheme to increase the data rate. These improvements are done dynamically on the frequency, allowing for the adaptation of maximum throughput based on the RF conditions. If there is interference on a portion of the band that is being utilized, the system scales back the throughput based on the conditions. This is accomplished by error-detection information in the delivered packets being analyzed at the receiving end for quality. If there is too high of an error rate, the receiving site responds to the sending location to modify the throughput to an acceptable level.

Third-Generation Networks

Migration of existing networks to 3G platforms will require major system upgrades. Because 3G is designed to operate on a broader spectrum, most base-station equipment will not be compatible with the standard. Because of this 2.5G systems have forestalled the upgrades by providing overlay data networks to the existing 2G infrastructure. The International Telecommunications Union's (ITU) has defined 3G within the International Mobile Telecommunications year 2000 (IMT–2000) initiative. Some key points made are as follows:

- PSTN voice quality

- High data throughput rates for mobile and stationary users (up to 2.048 Mbps for the later)

- Circuit and packet switch support for data services

- More efficient use of the existing spectrum

The prevailing standard that is driving 3G deployments is CDMA. There are varying forms of CDMA that are being deployed for 3G applications; however they all have similar architectures. Throughput for the next generation of CDMA is based on 5 MHz channels, which allow for a greater throughput in voice and data applications. Because of the higher throughput capabilities, a number of logical connections can be controlled by the end user, with varying data rates being applied. This can give the user the ability to have a mobile voice conversation of high quality while retrieving data for a Web page. This form of CDMA is known as wideband CDMA (WCDMA). Another acronym that is applied by the cellular industry to 3G systems is universal mobile telecommunications service (UMTS). The *wide* refers to the migration from the use of 1.25 MHz channels in narrowband CDMA to the 5 MHz channels. The capacity increase that is predicted for WCDMA is as least six times greater than GSM.

cdma2000

cdma2000 is a 3G technology designed to be backwardly compatible with the narrowband 1.25 MHz channels of existing 2G CDMA deployments. The data throughput of cdma2000 1X (the *x* represents the multimode carrier capability of the standard) is 144 kbps. More advanced data standards applied to cdma2000 are based on allocating dedicated data frequencies as base stations. This migration is termed *1xEV* for *evolution to DO* (which stands for data only) or *DV* (data and voice). 1xEV–DO has been deployed by a number of the major providers and tested throughput rates as high as 500 kbps. Bonding together three narrowband (1.25 MHz) channels together so they act as a parallel transmission pipe for data is the goal for the current configuration of CDMA in cdma2000. Data rates are expected to exceed 2 Mbps, similar to WCDMA without the cost of expensive migration at the base-station equipment end.

Time Division Synchronous Code Division Multiple Access (TD SCDMA)

It is important to note one final technology that is making its presence on the global marketplace. TD SCDMA is a 3G technology standard for the physical layer that is being championed by China Academy of Telecommunications Technology (CATT). Because China is the largest cellular market in the world (estimated 400 million users by 2007), the Chinese are trying to define the next generation of protocols to be used with cellular systems not only within their own country, but also on a global perspective. Siemens Corporation is a partner in developing this technology.

A key component differentiating TD SCDMA from other forms of CDMA is the use of time division duplexing (TDD) on the uplink and downlink from a user device. TDD shares the same frequency space rather than requiring a user to have two separate RF channels for communicating back to the base station. Synchronous time slots are allocated to provide efficient use of the frequency between the end points. When data applications are used across the TD SCDMA, asynchronous use of the time slots can be dynamically assigned to provide a greater amount of throughput for the downlink connection. Current 3G CDMA systems use frequency division duplexing (FDD), which requires a separate channel to perform these duties. *Figures 4.6* and *4.7* show the frequency and time slot allocations between TDD and FDD. Another key difference between TD SCDMA and other forms of CDMA is that it is defined as being deployed with smart antennas. Smart antennas are next-generation devices that utilize intelligent processing to narrowly define the RF signal to each user, which allows for greater frequency reuse within a cell. *Figure 4.8* depicts the difference between a regular cellular broadcast signal and a smart antenna signal.

Conclusion

With the various technologies that have been used to deliver cellular servic-es, networks have been continually improved to provide greater efficiency in respect to the licensed spectrum that is available. Competing global tech-nologies will limit the availability of a single, standardized platform that will allow for users to communicate with their wireless devices on different con-tinents.

Chapter 5: Antennas and Signal Propagation

One of the most easily recognizable components of any wireless system also happens to be a component that is more likely to vary from system to system. Antennas are all around us in the 21st century. An individual does not have to look hard to see towers along the side of the road with cellular or microwave antennas; the vast majority of automobiles have at least one antenna—possibly two if there are satellite radios or other specialty wireless devices in the car; and, of course, cellular phones also have antennas. Antenna characteristics, however, differ for each of these systems. Properties such as gain, radiation pattern, and frequency (among others) all play a part in the selection and performance of an antenna. In this chapter, we will cover these and other characteristics, and we will also discuss radio signal propagation through free space between antennas.

Antenna Performance Characteristics

Antennas are the devices in wireless systems that are responsible for focusing RF energy from an electrical interface (such as a radio transceiver) and radiating it into free space (in other words, air). An antenna, however, should be able to also gather radiated energy from free space and convert it back to an electrical signal for a radio device. This characteristic ability to transmit and receive equally well at a particular frequency is known as reciprocity.

Reciprocity is important for antennas because if this property were not true, two way communication systems would require two antennas — one for transmitting and one for receiving.

Note that reciprocity relies on the fact that the antenna transmits and receives at the same frequency. Although antennas may transmit or receive at wide ranges of frequencies, they will usually be optimized for a small set of frequencies. In cases where transmission and reception occur at different frequencies (such as in cellular environments), more than one antenna may be used to optimize communication in both directions.

Depending on the physical direction of an antenna, the RF energy it radiates will have a different polarization. That is to say, the direction of the electric field produced by the antenna depends on the attitude of the antenna. To obtain optimal communication, polarization between devices should be identical. Although cross-polarized systems may be able to communicate with one another, polarization differences will diminish the received signal strength between devices. (In fact, in some cases, wireless planners are able to eliminate interference between two links by implementing horizontal polarization on one and vertical polarization on the other.)

Gain and Radiation Patterns

Recall that RFs comprise just a portion of the EMS. Similarly, light resides on the EMS, although light waves are much shorter and have considerably higher frequencies than RF. Despite the differences between light and RF — and there are differences — some comparisons between these portions of the spectrum will aid in understanding certain properties of antennas. One of these properties is gain.

Gain is the ability of an antenna to focus RF energy and, in essence, amplify the RF signal. Reciprocity applies here, so gain affects both the transmission and reception of radio frequencies. Considering our light analogy, we can compare antenna gain in the transmit direction to flashlight lenses and, in the receive direction, lenses in prescription glasses. You have probably seen the difference between an ordinary flashlight with a cheap plastic lens and a similar flashlight with a glass lens. The glass lens is better suited to focusing the light energy, and therefore the light is visible over longer distances. This is identical to antenna gain in the transmit direction—various antennas have different focusing abilities, so they will transmit signals over different distances. Conversely, in the receive direction, antenna gain is similar to prescription eye lenses. Again, just as there are various prescription levels, there are various gain levels that an antenna may have. Gain requirements for an antenna rely on the ability of the connected radio device to decode signals at low levels.

Any change in the power of a signal is measured in units called decibels (dB). A gain in power is represented as an addition of some number of decibels, and a loss is suggested by the subtraction of some number of decibels. Note that a decibel only refers to the change in some power level—the unit, dB, does not imply an exact power level unless it is referenced to some known quantity. For example, dBm relates to a number of decibels, refer-

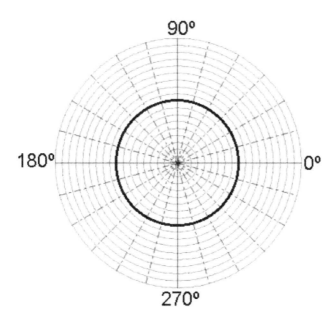

Figure 5.1: Isotropic Radiation Pattern

enced to 1 milliwatt of power—*dB* means nothing by itself, but it does when used in conjunction with the known power level *m*. Though seen less frequently than dBm in telecommunications, dBW (referenced to 1 Watt) is also used on occasion in radio systems. To add a bit more complexity to the subject, decibels are measured on a logarithmic scale instead of a linear scale. That is to say, 2 dBm is not twice as powerful as 1 dBm, as intuition would suggest. In reality an addition of 3 dB represents twice the original power, and a subtraction of 3 dB represents half the original power. To illustrate 23 dBm is twice as powerful as 20 dBm, and 17 dBm is half as powerful as 20 dBm.

Often different types of devices have different dB reference points. In radio transceivers, gain is most often described in dBm, as explained in previous paragraphs. Antenna specifications, however, generally refer to dBi, or decibels referenced to an isotropic source. An isotropic antenna is one that transmits and receives equally well in all directions, in all three dimensions. In practice, there is no such thing as an isotropic antenna because no antenna can be manufactured to transmit exactly as well in all directions. It is a use-

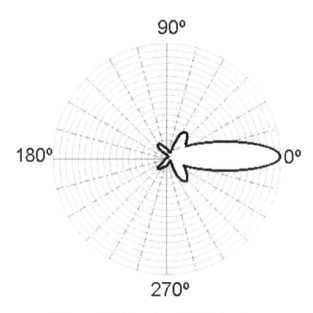

Figure 5.2: Directional Radiation Pattern

ful reference point, however, for discussing gain characteristics of real-world antennas. To understand how, compare *Figures 5.1* and *5.2*.

To acclimate yourself to these figures, imagine that you are standing at 0 degrees, looking directly at the antenna. This angle is generally considered the front of the antenna. From this point of view, 90 degrees would be the right side of the antenna, 270 degrees would be the left side, and 180 degrees would be the back of the antenna. Lines that are farther from the source represent higher gains and lines that are closer to the source suggest lower gains. Notice in *Figure 5.1* that the dark circle is the same distance from the source at all angles; thus, this is the radiation pattern of an isotropic source.

Figure 5.2 illustrates a directional antenna, which is one that transmits much better in a particular direction than in others. Notice, in comparing this figure to the previous one, that the antenna has sacrificed transmission power in other directions in order to increase its gain toward the front (0 degrees). To calculate the dBi value for this antenna, we compare its actual power directionality to the power it would produce if it were perfectly isotropic.

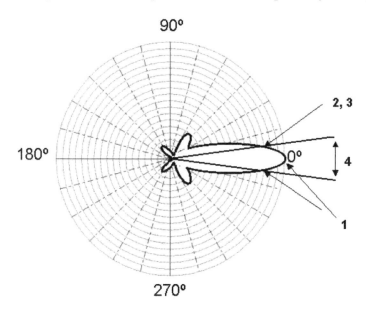

Figure 5.3: *Finding Beamwidth*

Assuming *Figure 5.1* represents the theoretical isotropic radiation pattern of the same source as *5.2*, we can easily calculate the gain of this antenna from various angles. At 0 degrees, we count 9 concentric circular lines between the isotropic graph and the directional graph. This means that the gain at 0 degrees is 9dBi, or 8 times the reference isotropic power. On the other hand, the sidelobes at approximately 60 degrees and 300 degrees have gains of –4dBi. In other words, their power levels are less than half of similar angles, assuming isotropic radiation.

Another performance characteristic related to the directionality of antennas is beamwidth. As we have seen, the maximum gain of a directional antenna is at 0 degrees. Beamwidth refers to the points on the radiation pattern where half of the maximum gain is achieved. Determining beamwidth on a radiation pattern chart takes four short steps: 1) Find the single point of highest gain; 2) Trace the radiation pattern back toward the source in both directions until you arrive 3dB back from the highest gain; 3) Draw lines from the source to both of these half-power points; and, 4) Measure the angle of the

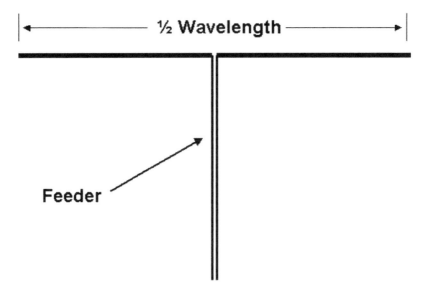

Figure 5.4: Dipole Antenna

straight lines—this angle is the beamwidth. To illustrate these steps refer to *Figure 5.3*.

Antenna properties such as reciprocity, polarity, gain, radiation pattern, and beamwidth influence how wireless systems are designed. With a thorough understanding of these characteristics as well as the available types of antennas, wireless application designers are well on their way to planning their links. As such, let us now turn our attention to types of antennas.

Types of Antennas

There is a plethora of types of antennas, many of which have interesting names like *dipole*, *horn*, and, everyone's favorite, *Yagi uda*. A discussion of all types of antennas is beyond the scope of this text, as many types serve special purposes that have little significance in cellular networks. Three antenna types, however, are worth discussing in some detail: dipoles, parabolic reflectors, and smart antennas.

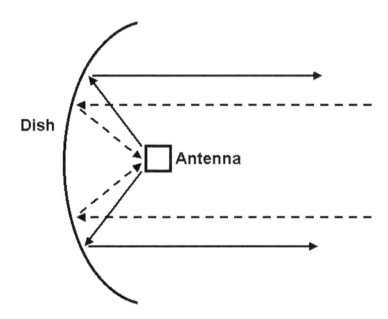

Figure 5.5: *Parabolic Reflector*

Dipole antennas are among the most common in many different wireless environments. This type of antenna consists of an RF feeder, such as a coaxial cable connected to a radio and two poles (see *Figure 5.4*). In length, the antenna is almost always some odd multiple of the primary frequency that it will transmit and receive. Antenna theory suggests that this method of determining length will produce the most accurate transmission and reception. Dipoles are considered omnidirectional; they transmit and receive well in all directions on the horizontal plane but do not offer extraordinary gain in any direction.

One of the best examples of a highly directional antenna is a parabolic reflector or parabolic dish. Here, the antenna is cradled by a reflective dish, which adds powerful focus to transmitted radio waves and also gathers and reflects received RF waves to the antenna. *Figure 5.5* shows a cross section of a parabolic reflector. Note that the parabolic dish surrounds the actual antenna (the arm that physically connects the antenna to the dish is not shown). Parabolic reflectors are always highly directional, but gain is directly proportional to the size of the dish. For instance, a 6-foot dish provides higher gain than a 3-foot dish. Parabolic reflectors are often found on the sides of cell towers, but they are not used for cellular communication purposes—a highly directional parabolic reflector is not useful for communicating with mobile devices. Instead, these dishes are used for microwave links between cell towers. Remember from *Chapter 2* that each BTS, or cell site, must have a connection back to the MTSO. Instead of putting in copper or fiber trunks from every BTS to the nearest MTSO, it is much more cost effective to connect cell sites wirelessly. Numerous companies produce radios for this exact purpose that operate in license-free portions of the spectrum, so cellular providers do not have to lease any more frequencies to accomplish this goal. Expensive copper or fiber trunks will still be required at certain points in the network, but this method of operation allows for considerably fewer trunks. Naturally fewer trunks yield fewer costs, and fewer costs allow for lower subscriber prices.

The last type of antenna we will discuss is relatively new. Smart antennas are BTS devices that dynamically adapt to varying radio conditions. One important possibility that this yields is the ability to track mobile users with a very narrow beamwidth signal. This allows providers to reuse frequencies, timeslots (in TDMA systems), and codes (in CDMA systems) numerous times within the same cell. Because these extremely narrow beamwidths separate users by space, this process is sometimes referred to as space division multiple access (SDMA). Reports indicate SDMA has increased TDMA capacity up to three times within cells; CDMA capacity has reportedly increased five times within cells. (CDMA will be discussed further in *Chapter 6*).

Numerous antenna types exist for an equal number of purposes. Dipoles, parabolic reflectors, and smart antennas represent a few of the most common types of antennas in use in cellular networks today. With a firm understanding of antenna principles and types, we can explore concepts of signal propagation between antennas.

Signal Propagation

The movement of radio waves through space is called propagation. The way in which radio waves propagate is a function of the frequency of the wave. In general, there are three methods of signal propagation: ground-wave propagation, sky-wave propagation, and line of sight (LoS) propagation.

At the low end of the RF spectrum, signals travel from antenna to antenna via ground-wave propagation. Radio frequencies at approximately or below 2MHz follow the contour of the Earth. As such, antennas in this frequency range do not need a clear line of sight between them in order to communicate. In fact, these signals will often propagate for many miles before finally attenuating so much that they are no longer useful. This explains why we can frequently tune into our favorite AM radio station even when we are very far away from the nearest transmitter.

Not all radio frequencies travel along the ground—slightly higher frequencies, such as those associated with CB and amateur radio, travel via sky-wave propagation. These frequencies are high enough to move away from the ground, but they cannot escape through the Earth's atmosphere. Once

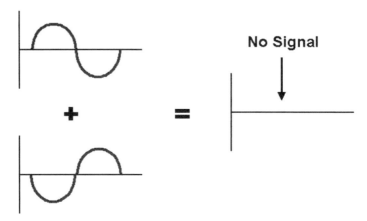

Figure 5.6: Destructive Interference Caused by Multipath

these waves reach the ionosphere, they are reflected back toward the ground. From the ground, these signals will commonly be reflected toward the ionosphere again. This continual bouncing allows these frequencies to travel long distances, like signals that propagate over the ground. On average, however, these distances are not as far as those of ground wave propagation.

The final form of propagation is LoS propagation. As the name suggests, this propagation type requires a visible, obstruction-free line from one antenna to the next. Frequencies over approximately 30 MHz all require LoS, and the necessity of LoS increases along with frequency. For instance, near LoS may suffice for communications at 30 MHz, but microwave transmissions at, say, 10Ghz, must absolutely have a clear path. Because cellular signals operate above 800 MHz, LoS is important to consider in cellular networks. Luckily mobile devices' antennas are omnidirectional, so they can pick up reflections of cellular signals even if they do not have clear LoS to the nearest tower. Of course, there is no guarantee that these devices will be able to access network services in buildings or other structures that prevent reflected signals from reaching a user. Also it is easy to recognize that these signals do not travel as far as those that are subject to ground-wave or sky-wave propagation. Cell towers are usually only a few miles from one another, which means that each cell site is only capable of serving a small area. As mobile devices move relatively short distances from cell towers, LoS drops quickly and communication soon fails. Thus, although true LoS is rarely available in a mobile environment, the frequencies in use dictate that mobile devices must be close enough to LoS that they can still receive reflected energy from each other.

RF Propagation Impairments

A number of potential propagation issues present themselves as RF signals travel through air or space. Attenuation, free space loss, noise, and multipath are a handful of the most common problems that affect wireless communications. Unlike wired communication, where some individual has complete control over transmission facilities, such potential impairments cannot be precisely controlled in wireless environments. As a result, wireless devices and applications must be designed to operate properly even in the presence of these potential impairments.

Perhaps the biggest potential impairment to wireless communication is attenuation or a reduction in signal strength. Gain is the opposite of attenuation—whereas gain increases signal strength, attenuation reduces it. Although signals suffer some attenuation in transit between radios and their antennas, the vast majority of signal-strength reduction is in between antennas. The phenomenon of drastic power loss in the air between antennas is known as free space loss. Without an understanding of approximately how much signal fading will occur in free space, a wireless system will fail!

Noise is another important factor to consider when examining wireless applications. Although there are countless sources of noise, it is generally considered any unintended or undesirable energy that degrades the quality or performance of a communication signal. In wireless systems noise commonly comes from both natural and manmade sources. For instance lightning generates sudden spikes of energy, which can affect wireless transmissions—these spikes are called impulse noise. In addition if two parties in an area attempt to transmit on the same radio frequency, their signals will collide and create noise as well. Regardless of the planning and quality of a communication system, noise is unavoidable. In fact even the electrons moving inside of devices create a phenomenon known as thermal noise. Because some level of noise is unavoidable, wireless engineers often refer to noise floors as the minimum acceptable radio reception levels. For radios to be able to detect information in a narrowband signal properly, the signal must be above this level. Otherwise the receiver will not be able to recover the signal from incidental noise.

Just as radio frequencies can reflect off the ground or the ionosphere, it can also bounce off of buildings, mountains, water, and many other objects. In cellular networks, this can cause issues when RF energy takes more than one path from a transmitter to a receiver, due to reflecting off of objects between the transceivers. This phenomenon, called multipath, becomes especially problematic when one path is longer than the other because it leads to destructive interference. *Figure 5.6* illustrates the effects of multipath where the same signal arrives at a receiver 180 degrees out of phase. Notice that the receiver combines the signals, and, in doing so, completely loses the signal. Of course, this is a worst-case scenario—in reality, signals rarely arrive 180 degrees out of phase. Any phase difference between received signals, however, will lead to some signal cancellation and will therefore make communication less efficient.

Conclusion

In this chapter, we have discussed a number of issues related to antennas and signal propagation. Although they are perhaps the most easily recognized elements of a wireless system, antennas are not easy to understand fully. To appreciate wireless systems, one must comprehend basic properties of antennas. In addition, wireless network planners should understand characteristics of different types of antennas to plan their applications properly. Would omnidirectional antennas serve the application best or should directional devices be used? How much gain should the antenna provide? These are the kinds of questions that designers must consider. Likewise signal propagation characteristics and potential impairments should be considered. Will the system operate at frequencies that will travel via ground-wave, sky-wave, or LoS propagation? Where are the key sources of noise in the system, and can they be eliminated or reduced? Again these questions are critical to design-ing a wireless deployment successfully.

Up to this point we have focused on narrowband transmission techniques. Although narrowband signals seem to make the most efficient use of the spectrum, we will find out in the next chapter that system capacity can be increased with wideband techniques known as spread spectrum.

Chapter 6: Spread Spectrum in Cellular

Conventionally a radio signal is transmitted at a particular frequency or within a small range of frequencies at a relatively high power. This narrowband method of operation, however, gives rise to two potential disadvantages: third parties can either intentionally or unintentionally interfere with this communication by transmitting another signal at the same frequencies that are in use, and third parties can easily eavesdrop on the frequencies in use without permission. To combat these pitfalls, spread spectrum (SS) techniques may be used. As the name suggests, the essence of these techniques is to distribute an information signal over a wider range of the available spectrum—in other words, instead of a narrowband transmission, wideband operation is preferable in SS systems. *Figure 6.1* is an example of a narrowband signal—notice that the transmitted signal is heavily concentrated near the carrier frequency, with power dropping sharply on either side of the carrier. On the other hand a wideband signal remains strong much farther from the carrier, as illustrated in *Figure 6.2*.

With such a premium on RF spectrum, it may seem counter-intuitive that SS systems actually use more spectrum than straightforward modulation. There, however, is one other key component to spread spectrum systems: a pseudorandom code or pattern is used to spread the signal in such a way as to appear

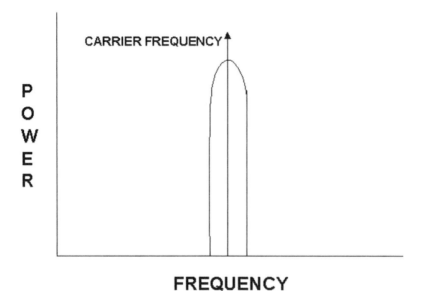

Figure 6.1: Narrowband Signal

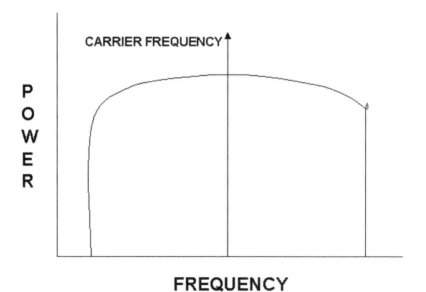

Figure 6.2: Wideband Signal

as noise to any device that does not know the spreading code or pattern. Although different types of spread spectrum techniques use these pseudo-noise (PN) codes differently, these codes are imperative to SS systems. Because of this it turns out that many devices can actually use the same part of the spectrum at once. This is why spread spectrum implies that it is better to use wideband transmission than narrowband transmission.

Spectrum techniques can offer huge advantages to technologies that operate in license-free portions of the radio spectrum. The vast majority of the spectrum is controlled by the government, and organizations that wish to use any portion of it must lease that portion from the government. The U.S. FCC, as well as other governments' communications regulatory bodies, has set aside certain portions of the spectrum for common use. As such, many technologies and vendors use these license-free portions of the spectrum—after all no one wants to pay a huge lease fee to set up a wireless access point for a home network. The downside here is that interference will become an ever-greater issue as these portions of the spectrum become saturated. To allow the largest number of users to be able to access these frequencies without interfering with one another, spread spectrum makes sense. This allows multiple parties to use the same frequency bands as one another and still filter out others' conversations. In fact, until recently, the FCC required technologies operating in unlicensed bands to use spread spectrum. Various vendors, however, developed alternatives to SS, and the FCC has acknowledged that these techniques will suffice, as long as they prevent excessive interference.

Two forms of spread spectrum are commonly used today in wireless systems: frequency hopping spread spectrum and direct sequence spread spectrum. We will examine both of these in order.

Frequency Hopping Spread Spectrum

Frequency hopping spread spectrum (FHSS) uses a shifting carrier frequency for data transmission. Instead of transmitting all data on a given carrier frequency for the complete duration of communication, a pseudo-random code is used to jump between different carrier frequencies after a given period of time. *Figure 6.3* illustrates the relationship between transmission fre-

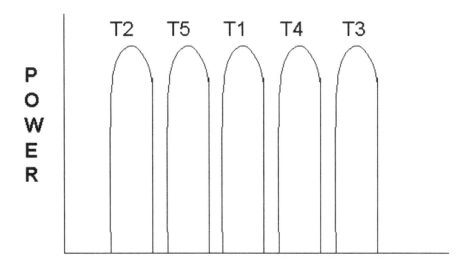

Figure 6.3: FHSS (Frequency vs. Power)

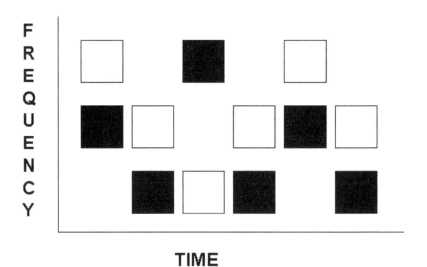

Figure 6.4: FHSS (Time vs. Frequency)

quency, time, and power in FHSS systems. For a period of time, T1, the carrier frequency in this example, is the middle frequency. Notice that for this time, the signal will be transmitted at the same full power as a typical narrowband application would require. If we were to take a snapshot at this time, we would essentially have a typical narrowband communication channel. After a specified time known as the dwell time, however, notice that the carrier frequency changes. The figure indicates this as the leftmost frequency, denoted by time period T2. The T2 dwell time will be identical to that of T1. Upon completion of the dwell time, the carrier frequency will shift to the frequency indicated by time period T3. This pattern would continue in a seemingly random fashion through T4, T5, and so on.

Figure 6.4 presents another graphical way to consider FHSS where there are multiple devices communicating in the same wideband frequency space. Consider the white boxes to be one conversation and black boxes, a second. Notice that each conversation hops from frequency to frequency after a specified dwell time. Statistically, it is probable that the conversations will not jump to the same carrier frequency at the same time. There are three carriers in this example and only two conversations. As such, there is only a one-third chance that both conversations will land on the same carrier frequency at the same time. (Of course, in reality, many more carriers would be used, so that the probability of collisions would be even lower.) Note that when collisions do occur, the colliding conversations will suffer for just a single dwell time.

Of course if two devices must communicate via FHSS, both must know the pseudo-random code in order to properly switch between carrier frequencies. In addition, perfect synchronization is required between devices, such that they will also anticipate when the move to the next carrier frequency. The second point here—synchronization—adds to the complexity of frequency-hopping devices.

FHSS is used more often in wireless LAN environments than in cellular networks. Cellular networks use DSSS and a closely related concept, CDMA, much more often than FHSS, and we will discuss these technologies next.

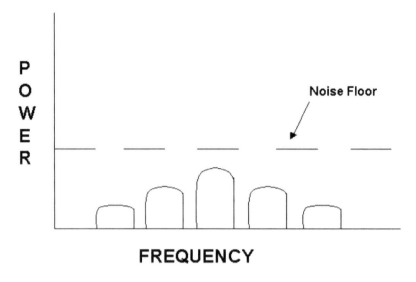

FREQUENCY

Figure 6.5: DSSS

DSSS

Where FHSS hops between carrier frequencies to spread a narrowband signal, DSSS uses a number of frequencies concurrently. Instead of transmitting at full power, however, each frequency is given a small output level. In fact DSSS transmissions may exist below the so-called noise floor, meaning that any device that does not know how to despread the signal will simply disregard the transmission as noise. *Figure 6.5* illustrates this concept.

A PN code is used to spread the information signal across all of the frequencies. In DSSS systems this PN code is referred to as the chipping code or, alternatively, the spreading code. The chipping code is a series of bits that will be multiplied into actual information bits; to be effective, the chipping rate must be significantly greater than the bit rate of the information signal. Performance of a DSSS system is directly related to the chipping rate of the spreading code. In other words as the processing gain (ratio of chipping rate to information rate) increases, so does the system's immunity to interference. *Figure 6.6* illustrates the combination of an information bit and the spreading code. This combination is accomplished via an exclusive OR, whereby a

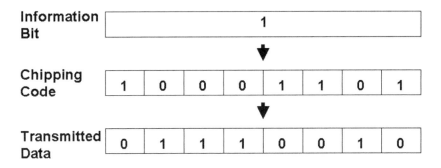

Figure 6.6: Combination of Data and Chipping Code at the Transmitter

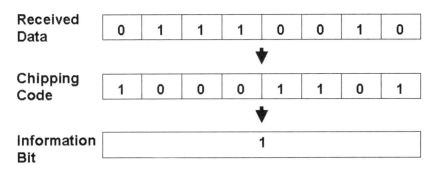

Figure 6.7: Combination of Received Signal and Chipping Code at the Receiver

combination of 0+0=0, 1+1=0, and 1+0 or 0+1=1. The output of this exclusive OR is transmitted on the various frequencies within the DSSS at the same time. At the receiver each of these frequencies are examined and the output is again exclusive ORed with the PN spreading code. The output of this combination should yield the original signal. *Figure 6.7* indicates the logic at the receiver. If any errors have occurred on any of the composite frequencies, the result of the combination of received signal and chipping code is put through a "majority rules" process. Essentially, if the result looks more like a 1, the receiver assumes it to be a 1 and vice versa.

To explore direct sequence techniques fully requires a bit of an excursion in mathematics and electrical engineering, which we will not cover here. The

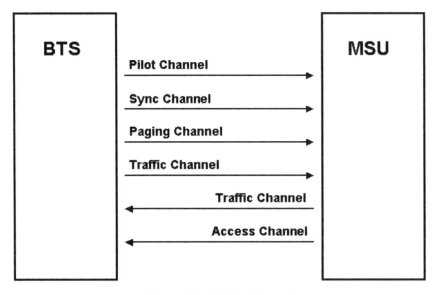

Figure 6.8: CDMA Channels

key to understanding DSSS, however, is to realize that an individual information bit may be transmitted in multiple frequencies over a wide bandwidth, as a result of the rapid PN code. DSSS is used in numerous wireless technologies, and we will spend some time discussing the one most relevant to cellular networks: CDMA.

CDMA

Whereas DSSS is used in a number of other wireless applications, CDMA is a cellular technology that is built upon direct sequence techniques. (Note: Some synonymous terms for CDMA include IS–95 and ANSI–95, which are standards that define the technology.) To understand CDMA, consider its alternatives, which were discussed briefly in *Chapter 3*, FDMA and TDMA. As you will recall, FDMA and TDMA are both designed to allow as many users as possible to access the same medium, the RF spectrum, in order to communicate. FDMA specifies that a user may have total control of a frequency for the duration of a conversation. TDMA builds upon this by allowing users to access the same frequency at regularly recurring time intervals — this increases the capacity of the system. CDMA uses DSSS to allow users to access the same frequency at the same time as other users, which, again, increases capacity.

CDMA Channel Types

There are five different channel types in CDMA systems, which are used for various overhead purposes in addition to voice and data transmission. When discussing CDMA, reference is often made link direction. The forward link refers to any communication from a BTS to an MSU; conversely, the reverse link refers to transmission from a MSU to a BTS. On the forward link the relevant channel types are pilot channels, sync channels, paging channels, and traffic channels. On the reverse link, we find access channels and traffic channels. *Figure 6.8* summarizes the direction of these channels.

Pilot channels are transmitted by BTSs continuously to announce their presence to any potential mobile users. Because the pilot channel is the first signal for which a new mobile user will scan, the pilot channel code must be identical among all cell sites. Because all BTS units transmit on the same frequency and use the same pilot code, a function known as the pilot offset is used to identify a BTS. Identifying all nearby cell sites via unique pilot offsets allows a mobile user to determine which site it can get the strongest signal from, which, in turn, provides the best possible performance quality. The pilot channel accounts for 20 percent of the strength of transmission from the BTS. Because this is a standardized number, MSUs can derive how far they are from their BTS, based on the strength of the pilot channel. An MSU can then use this information to raise or lower its own transmit power to overcome the near-far problem, described in the DSSS section of this chapter, which we will discuss later. There is only one pilot channel per BTS.

A sync channel is transmitted continually from the BTS. Because synchronization is maintained within and between cells, MSU devices use the sync channel upon startup to lock onto this synchronization as well as determine the proper pilot offset for a BTS. Once a mobile device has locked onto the proper timing, it no longer needs the sync channel (unlike the pilot channel). As with the pilot channel, there is only one sync channel per BTS.

Paging channels allow a BTS to signal incoming calls as well as to provide certain overhead commands to a mobile user. In addition to signaling calls, traffic channel assignments are carried out on paging channels. Unlike the previous forward link channel types, each BTS has seven individual paging channels.

Traffic channels, as the name suggests, are used to pass voice and data. On the forward link there are a total of 55 available traffic channels per BTS. Once the initial call has been signaled on either the paging or access channels, depending on link direction, call control travels on traffic channels in addition to the voice or data traffic.

The only channel type unique to the reverse link is the access channel. This channel type serves the same purpose as the paging channel in the forward direction: its primary purpose is to signal outgoing calls. In addition, however, it may be used to respond to pages from a BTS that are unrelated to call control.

Soft Handoff

In addition to greater system capacity, CDMA provides the ability to perform soft handoffs between cells. When a mobile user moves from cell to cell, each cell site will require the mobile device to switch to a frequency or time slot that differs from what was used in the previous cell. This process requires a brief loss of service, while the mobile device switches to accommodate the requirements of the new cell. Because of the brief loss of service, this is known as a hard handoff.

On the other hand, CDMA systems do not require such a change because all sites use the same frequencies and timeslots are not defined. This allows a mobile device to establish communication with more than one BTS at a time. So while a mobile user is communicating to the PSTN through BTS 1, a mobile device can already be in the process of determining a PN code to use on BTS 2. Once it has done this, it can seamlessly switch over to BTS 2 without the user knowing the handoff has occurred. The grace with which this process is handled is why it is known as a soft handoff.

Near/Far Problem

CDMA offers a number of advantages over its frequency and time division counterparts, but it suffers from one great problem. The so-called near/far problem arises from the fact that many mobile users could potentially communicate to a BTS on a single frequency at various distances. If all such devices transmitted at the same power level, transmissions from those MSUs farther from the BTS would attenuate more than those from MSUs near the BTS. As such, the MSUs nearest the BTS would overpower those farther away, and the mobile devices that were farther away would not be served properly.

To combat this, mobile devices are able to adjust their own transmit power levels based upon their distance from a BTS. Recall that the pilot channel on every BTS must remain at a certain power (20 percent of the total transmit power). Mobile devices are programmed such that they can use the received pilot channel strength, compared with the theoretical pilot channel strength to determine their distance from a particular BTS. Once they have calculated this distance, they can then adjust their transmit power up or down to alleviate the near/far problem.

Conclusion

Spread spectrum techniques allow wireless systems to increase the capacity of potential users, while at the same time combating both intentional and unintentional interference. Surprisingly this capacity increase is accomplished by using wideband signals, which at first glance would appear to waste spectrum resources. The two most common spreading techniques include frequency hopping spread spectrum and direct sequence spread spectrum. In cellular environments, direct sequence techniques are used more often than their frequency-hopping counterparts, especially when used in conjunction with CDMA. CDMA is a cellular-access technology that improves the potential capacity in a cellular network over FDMA and TDMA. In addition to the DSSS–based access techniques, CDMA also defines a number of specific channels that are used for both overhead purposes as well as user traffic between BTSs and MSUs.

References

[1] Wikipedia, The Free Encyclopedia. Retrieved September 7, 2005 from http://en.wikipedia.org/wiki/PSTN.

[2] Cell Phone Glossary — Mobiledia. Retrieved September 7, 2005 from http://www.mobiledia.com/glossary/166.html.

[3] IT Paper. Retrieved September 9, 2005 from http://www.telect.com/publications/pdfs/starwrls.pdf.

[4] Encyclopedia article about cell site. Retrieved September 7, 2005 from http://encyclopedia.thefreedictionary.com/Cell+site.

[5] Jones, S. and Kovac, R. (2003). *Introduction to Communication Technologies: A Guide for Non-Engineers*. Auerbach Publishing: Boca Raton, FL.

[6] Wisniewski, S. (2005). *Wireless and Cellular Networks*. Pearson Prentice Hall: Upper Saddle River, NJ.

[7] Stallings, W. (2002). *Wireless Communications and Networking*. Prentice Hall: Upper Saddle River, NJ.

[8] Verizon Learning Center. Retrieved September 8, 2005 from http://www22.verizon.com/about/community/learningcenter/articles/displayarticle1/0,,1008z1,00.html.

[9] Illinois Institute of Technology. Retrieved September 8, 2005 from http://ir.iit.edu/publications/downloads/iitr-Car_journal.pdf#search= 'reuse%20frequency.

[10] "Call Drop." Retrieved September 11, 2005 from http://www.moto-zone.com.au/motoglossary/motoglossary.asp#CallDrop.

[11] "An Efficiency Limit of Cellular Mobile Systems." Retrieved September 15, 2005 from http://www.ctr.columbia.edu/~campbell/andrew/publications/papers/mmt98.pdf.

[12] "Adaptive and Predictive Downlink Resource Management in Next Generation CDMA Networks." Retrieved September 15, 2005 from http://www.bell-labs.com/user/ramjee/papers/info04cdma.pdf.

[13] "TechOnline—Education Resources for Electronic Engineers." Retrieved September 15, 2005 from http://www.techonline.com/community/ ed_resource/feature_article/14863. 9/15/05

[14] "Reverse Link Power Control." Retrieved September 18, 2005 from

http://www.qualcomm.com/ProdTech/cdma/training/cdma25/m6/m6p0 9.html.

[15] Miller, G. M. and Beasley, J. S. (2002). *Modern Electronic Communication*. 7th edition. Prentice Hall: Upper Saddle River, NJ.

[16] Rappaport, T. S. (2002). *Wireless Communications: Principles and Practices*. 2nd edition. Prentice Hall: Upper Saddle River, NJ.

[17] Bedell, P. (2001). *Wireless Crash Course*. McGraw-Hill: New York.

[18] GSM World. (n.d.) Retrieved September 9, 2005 from http://www.gsmworld.com/about/history/.

[19] Privateline. (n.d.) Retrieved September 9, 2005 from http://www.privateline.com/Cellbasics/Cellbasics.html.

[20] Radio Electronics. (n.d.) Retrieved September 9, 2005 from http://www.radio-electronics.com/info/cellulartelecomms/pdc/ pdc-summary.php.

[21] Wikipedia, (n.d.). "W–CDMA." Retrieved Sep. 09, 2005, from http://en.wikipedia.org/wiki/WCDMA.

[22] RAD data communication. (2000). "What is GSM? Global System for Mobile Communications." Retrieved September 28, 2005 from http://www.pulsewan.com/data101/gsm_basics.htm.

[23] Wood, Lloyd. (2000). "GSM Overview." Retrieved September 28, 2005 from http://www.ee.surrey.ac.uk/Personal/L.Wood/constellations/tables/gsm.html.

[24] Federal Communication Commission. (n.d.). "Wireless Communications Service." Retrieved September 28, 2005 from http://wireless.fcc.gov/wcs/.

[25] International Engineering Consortium, (n.d.). "Mobile Telephone System Using the Cellular Concept." Retrieved September 12, 2005, from http://www.iec.org/online/tutorials/cell_comm/topic02.html? Next.x=39&Next.y=30.

[26] Unstrung, (n.d.). "First Generation (1g)." Retrieved September 12, 2005, from http://www.unstrung.com/document.asp?doc_id=16857&page_number=2.

[27] CDMA Development Group. (n.d.). "The CDMA Revolution." Retrieved September 28, 2005 from http://www.cdg.org/technology/cdma_technology/ a_ross/cdmarevolution.asp.

[28] Simo, Ernest. (n.d.). "Introduction and Forward Link Power Control." Retrieved September 28, 2005 from http://www.cdmaonline.com/members/ 2ginteractive/3000/A26.htm.

[29] Punters, Steve. (2004). "CDMA vs. TDMA." Retrieved September 28, 2005 from http://www.arcx.com/sites/CDMAvsTDMA.htm.

[30] Beaulieu, Mark. (2002). "Understanding the Three Wireless Network Groups of the Internet." Retrieved September 28, 2005 from http://www.awprofessional.com/articles/article.asp?p=24904&seqNum=8

[31] Federal Communications Commission, (2002). "Operations." Retrieved September 12, 2005, from http://wireless.fcc.gov/services/cellular/operations/.

[32] Mobiledia, (n.d.). "Mobile Telephone Switching Office (MTSO)." Retrieved September 12, 2005, from http://www.mobiledia.com/glossary/166.html.

[33] Atallah, Jad. (2004). "Handover Considerations in the Design of Multi-Standard Transceiver Front Ends." Retrieved September 28, 2005 from http://64.233.167.104/search?q=cache:7BUHknPmJpMJ:www.it.kth.se/ courses/2G1330/Jad_Atallah_report_revised20040625.pdf+soft+handoff+design+considerations&hl=en.

[34] Audit My PC.com, (n.d.). "FDMA: Retrieved September 30, 2005, from http://www.auditmypc.com/acronym/FDMA.asp.

[35] Wikipedia, (n.d.). "Frequency-Division Multiple Access." Retrieved September 30, 2005, from http://en.wikipedia.org/wiki/Frequency-division_ multiple_access.

[36] Qualcomm. (n.d.). "Introduction to CDMA." Retrieved September 28, 2005 from http://www.qualcomm.com/ProdTech/cdma/training/cdma25/m6/m6p03.html

[37] Agathangelou, Marios C. (2003) "CDMA System Architecture." Retrieved September 28, 2005 from http://www.us.design-reuse.com/news/ ?id=6647&print=yes.

[38] Stüber, G. L. (2001). "Principles of Mobile Communication." 2nd edition. Kluwer Academic Publishers: AH Dordrecht, The Netherlands.

[39] TD SCDMA Forum (n.d.) Retrieved December 30, 2005 from http://www.tdscdma-forum.org/EN/resources/index.asp.

[40] http://kabuki.eecs.berkeley.edu/~rsn/papers/224termpap.pdf.

[41] http://www.bee.net/mhendry/vrml/library/cdma/cdma.htm.

[42] http://www.ieee-virtual-museum.org/collection/tech.php?id=2345914&lid=1.

[43] http://www.rod.beavon.clara.net/samuel.htm.

[44] http://www.morsehistoricsite.org/history/morse.html.

[45] http://sln.fi.edu/franklin/inventor/bell.html.

[46] Farley, T. and M. Van Der Hoek. "Cellular Telephone Basics: AMPS and Beyond." http://www.privateline.com/Cellbasics/Cellbasics.html http://nobelprize.org/physics/laureates/1909/marconi-bio.html.

[47] http://www.cellularone.com/IndustryOverview.asp.

[48] Benjamin, S. M., D. G. Lichtman, and H.A. Shelanski, (2001). *Telecommunications Law and Policy.* Carolian Academic Press: Durham, NC.